SQL Server on Azure Virtual Machines

A hands-on guide to provisioning Microsoft SQL Server on Azure VMs

Joey D'Antoni, Louis Davidson, Allan Hirt, John Martin, Anthony Nocentino, Tim Radney, and Randolph West

SQL Server on Azure Virtual Machines

Copyright © 2020 Packt Publishing

Authors: Joey D'Antoni, Louis Davidson, Allan Hirt, John Martin, Anthony Nocentino, Tim Radney, and Randolph West

Managing Editors: Aditya Datar and Siddhant Jain

Acquisitions Editor: Alicia Wooding

Production Editors: Ganesh Bhadwalkar and Deepak Chavan

Editorial Board: Vishal Bodwani, Ben Renow-Clarke, Ian Hough, and Dominic Shakeshaft

First Published: June 2020

Production Reference: 1010620

ISBN: 978-1-80020-459-1

Published by Packt Publishing Ltd.

Livery Place, 35 Livery Street

Birmingham B3 2PB, UK

Table of Contents

Chapter 4: SQL Server on Linux in Azure Virtual Machines　93

Chapter 7: Hybrid scenarios (Microsoft SQL IaaS)   153

Appendix A   171

Index   177

Foreword

SQL Server on Azure Virtual Machines combines the industry-leading performance and analytics of SQL Server with the security and flexibility of Azure. It is a common destination for lift-and-shift SQL migrations to the cloud while maintaining full SQL Server compatibility and operating system–level access.

The intention of this book is to be a technical guide for SQL Server on Azure Virtual Machines, Microsoft's **infrastructure as a service (IaaS)** offering for SQL Server. The book begins with an overview of Microsoft's Azure SQL family of SQL Server–related data services in the cloud. Tips for getting started and hero capabilities are discussed before deep dives into security, Linux, and performance. Finally, best practices are shared for cloud migrations and hybrid scenarios across on-premises and cloud environments.

Each chapter was written by Microsoft Data Platform MVPs and brings their unique perspective, with input from Microsoft Engineering subject matter experts.

Preface

About

This section briefly introduces the authors, the coverage of this book, the technical skills you'll need to get started, and the hardware and software requirements required to complete all of the included activities and exercises.

About SQL Server on Azure Virtual Machines

Deploying SQL Server on Azure virtual machines allows you to work on full versions of SQL Server in the cloud without having to maintain on-premises hardware.

The book begins by introducing you to the SQL portfolio in Azure and takes you through SQL Server IaaS scenarios, before explaining the factors that you need to consider while choosing an OS for SQL Server in Azure VMs. As you progress through the book, you'll explore different VM options and deployment choices for IaaS and understand platform availability, migration tools, and best practices in Azure. In later chapters, you'll learn how to configure storage to achieve optimized performance. Finally, you'll get to grips with the concept of Azure Hybrid Benefit and find out how you can use it to maximize the value of your existing on-premises SQL Server.

By the end of this book, you'll be proficient in administering SQL Server on Microsoft Azure and leveraging the tools required for its deployment.

About the Authors

Louis Davidson (Chapter 1, Introduction to SQL Server on Azure Virtual Machines)

Louis Davidson is a data architect for CBN in Virginia Beach, VA; telecommuting for many years from Cleveland, TN (which is not even as glamourous as it sounds.) Louis has written and contributed to many books on SQL Server topics over the past 20 years. His most prominent work has been five editions of his book entitled: "Pro SQL Server Relational Database Design and Implementation" for Apress in 2016, with a new version forthcoming in 2020. Louis has been a speaker at many SQL Saturday events, and has helped organize events in Nashville and Chattanooga, TN.

Allan Hirt (Chapter 2, Getting started with SQL Server on Azure Virtual Machines)

SQLHA, LLC founder, consultant, trainer, author, and business continuity, infrastructure, and virtualization expert Allan Hirt has been working with SQL Server since 1992 when it was still a Sybase product as well as clustering in Windows Server since the late 1990s when it was known as Wolfpack. Currently a dual Microsoft MVP (Data Platform; Cloud and Datacenter Management) as well as a VMware vExpert, Allan works with all sizes of customers no matter if they are on premises or in the public cloud and delivers training and speaks at events over the world.

Joey D'Antoni (Chapter 3, Hero capabilities of SQL Server on Azure Virtual Machines)

Joseph D'Antoni is a Principal Consultant at Denny Cherry and Associates Consulting. He is recognized as a VMWare vExpert and a Microsoft Data Platform MVP and has over 20 years of experience working in both Fortune 500 and smaller firms. He has worked extensively on database platforms and cloud technologies and has specific expertise in performance tuning, infrastructure, and disaster recovery.

Anthony Nocentino (Chapter 4, SQL Server on Linux in Azure Virtual Machines)

Anthony Nocentino is the Founder and President of Centino Systems as well as a Pluralsight author and a Microsoft Data Platform MVP, Linux Expert, and Corporate Problem Solver. Anthony designs solutions, deploys the technology, and provides expertise on system performance, architecture, and security. Anthony has a Bachelors and Masters in Computer Science with research publications in high performance/ low latency data access algorithms and spatial database systems. You can find him on Twitter **@nocentino**.

Tim Radney (Chapter 5, Performance)

Tim is a Data Platform MVP. He has presented at PASS, SQLintersection, Microsoft Ignite, SQL Saturdays, user groups and numerous webinars. In addition, Tim runs the Columbus GA SQL Users Group, is a PASS Regional Mentor and was named a PASS Outstanding Volunteer. He's married with three children and has a passion for electronics. He also farms chickens, crops, and tilapias in his spare time.

John Martin (Chapter 6, Moving workloads to SQL Server on Azure Virtual Machines)

John is an experienced data platform professional having spent over a decade working with the Microsoft data and cloud platform technologies. In this time John has learned how to get the most out of these platforms as well as the key pitfalls that should be avoided.

Randolph West (Chapter 7, Hybrid scenarios (Microsoft SQL IaaS))

Randolph West, founder of Born SQL, is an independent IT consultant, speaker, Calgary PASS user group leader, Microsoft Data Platform MVP, and lead author of the book *SQL Server 2019 Administration Inside Out*. Randolph specializes in SQL Server performance tuning, disaster recovery, and migrations from really old versions, with an emphasis on implementing best practices. Randolph has presented at PASS Summit, SQLBits, SQL Saturdays, and user groups. You can also find Randolph acting and directing on screen or the stage, or annoying people on Twitter. Do not trust Randolph around chocolate.

Learning Objectives

By the end of this book, you will be able to:

- Choose an operating system for SQL Server in Azure VMs
- Use the Azure Management Portal to facilitate the deployment process
- Verify connectivity and network latency in cloud
- Configure storage for optimal performance and connectivity
- Explore various disaster recovery options for SQL Server in Azure
- Optimize SQL Server on Linux
- Discover how to back up databases to a URL

Audience

SQL Server on Azure Virtual Machines is for you if you are a developer, data enthusiast, or anyone who wants to migrate SQL Server databases to Azure virtual machines. Basic familiarity with SQL Server and managed identities for Azure resources will be a plus.

Approach

This book incorporates every aspect of SQL deployment on Azure with a perfect blend of theory and hands-on coding. Each chapter is designed to build on the learnings of the previous lesson.

Hardware and Software Requirements

For the optimal experience, we recommend the following configuration:

For Windows:

- Processor: Minimum: x64 Processor: 1.4 GHz

    Recommended: 2.0 GHz or faster

- Memory: Minimum 2GB RAM
- Storage: 8 GB available space

For Linux:

- Processor: x64 Processor compatible only: 2 GHz 2 cores
- Memory: Minimum 2GB RAM
- Storage: 6 GB available space

Conventions

Code words in the text, database table names, folder names, filenames, file extensions, pathnames, dummy URLs, user input, and Twitter handles are shown as follows:

The following code shows how to generate a list of container images available for Red Hat Enterprise Linux (output abbreviated):

```
curl -sL https://mcr.microsoft.com/v2/mssql/rhel/server/tags/list
"2019-CU1-rhel-7.6"
"2019-CU1-rhel-8"
"2019-GA-rhel-7.0"
"2019-GA-rhel-7.6"
"2019-GDR1-rhel-7.0"
"2019-GDR1-rhel-7.6
"2019-latest"
"latest"
"vNext-CTP2.0"
```

> **Note**
>
> All the bitly links are listed at the end of each chapter.

1

Introduction to SQL Server on Azure Virtual Machines

By Louis Davidson

Introduction

In this chapter, we introduce the relational SQL Server products that you can use in Azure to store and process transactional data in a relational format, which is to say data stored in tables and columns. To get started on this discussion, there are a few terms and concepts that are important to understand.

The concepts of **platform as a service (PaaS)** and **infrastructure as a service (IaaS)** can be confusing and are used constantly when discussing services that allow you to build software on a cloud service. The fundamental distinction between the two lies in how **managed** the offering is. A service being managed means that the provider handles some amount of the operation (or management) of the service. When you install an operating system and SQL Server on your on-premises computer, you manage the entire hardware and software infrastructure yourself. This starts with making sure the server is plugged in and everything else moves from there. PaaS and IaaS both indicate **managed services** because the resources you use will be managed to some extent.

PaaS indicates that you are getting a platform to work with, and more of the management tasks such as software patching, performance tuning, backups, and fault tolerance will be handled by Azure. The goal is to let you focus on providing business value and leave the day-to-day operations to Microsoft. How much of the management is done for you is based on the features of the offering, but as an example, each of the PaaS offerings we will introduce will handle backups of your data without you or your customer thinking too much about it, until you find the need to restore your data.

The IaaS model primarily provides management of the hardware and network. You never need to, and never can, touch any of the physical resources or even access the location of the server. Just like when you get a new computer, an IaaS server may have software pre-installed for you, but once you take over the computer, managing and configuring the software and hardware is up to you. The Azure IaaS platform for SQL Server does include tools to help you automate the management of the software, giving you some of the characteristics of the PaaS model, but these tools will not be as controlled in the IaaS model as they will in a PaaS-model server.

A managed database service does not indicate that the Azure platform will change the meaning of any data or code you have written; in fact, it's quite the opposite. This is still part of what your organization needs to do. No changes will be made to your system that change the meaning of the structures you create. You will simply need to be less concerned with day-to-day processes that are common to pretty much every organization.

This book largely focuses on the IaaS offering using SQL Server on Azure Virtual Machines, though we will introduce the PaaS offerings for contrast. The rest of this chapter will introduce the Azure SQL offerings, the newest features in SQL Server 2019, and the value of using SQL Server in its IaaS configuration.

The Azure SQL portfolio

Microsoft Azure SQL is a modern SQL portfolio of offerings for storing relational data as a service. It is powered by the industry-leading SQL Server engine, which has evolved greatly over the years, retaining backward compatibility at the code level and continuing to provide monumental leaps in performance and storage capacity version over version. Some of the Azure SQL offerings are **evergreen**, meaning the offering is always up to date with the latest updates and patches. Because each offering is essentially based on the same SQL Server code, **database administrators (DBAs)** and developers can often use the tools and resources they are already familiar with from their past SQL Server experience, including graphical, command-line, and T-SQL-based tools, for much of the work they need to do.

For many organizations, having to build, house, and manage hardware and software is a large burden for a variety of reasons, but in most cases, cost and security are the most important. Beyond the easily quantified cost of purchasing hardware and software, there are costs in finding qualified persons to manage the hardware, **operating systems (OSes)**, and database platform, all before considering the day-to-day operations such as tuning queries and executing backups.

Using the Azure platform, upgrades to VM type and size can be done by a simple UI operation rather than buying new hardware, configuring it, and migrating all of your data onto it (that process may still occur, but your experience is often checking a box or sliding a slider on a web page and letting the automation do the work for you.)

The second reason is the most important: security. What all the data breaches in recent history have demonstrated is that most databases are accessible from the internet in some manner. Having the management and security of your infrastructure in the hands of a company such as Microsoft pushes the technology burdens of a very large part of securing your data onto them. You can take comfort in the fact that the entire Azure business model rests on the security of all its customers' data, including yours. It will still be your responsibility to build proper security principals with adequate passwords and two-factor authentication, as no security will stop a user with proper credentials from accessing your online resources.

Azure has multiple other database management systems in the Data Platform portfolio for different types of database needs, including Cosmos DB[1], and Synapse Analytics[2], to name a few. Here's a full list of current products in the Azure family of services[3]. Note that Synapse Analytics uses relational tables, but it is focused on large-scale, specialized analytics. This chapter focuses solely on the relational SQL Server–based offerings (Azure also has relational database offerings such as MySQL[4], PostgreSQL[5], and MariaDB[6]).

The Azure SQL portfolio provides a consistent and unified management experience spanning three SQL Server offerings in Azure, each with its own targeted use cases. Almost any of the offerings will be perfectly acceptable to provide support for transaction processing (commonly referred to as **online transaction processing (OLTP)**) as well as most analytics (reporting) scenarios. Each offering is positioned to provide different levels of service, as we will discuss. The three offerings we will discuss specifically are:

- SQL Server on Azure Virtual Machines.

- Azure SQL Managed Instance.

- Azure SQL Database.

> **Note**
>
> There is one additional method of deployment using containers that we will not be specifically covering in this book. The container method is very similar in functionality to the VM deployment, except the VM is replaced by a lightweight, GUI-free container running Linux or Windows using Kubernetes or Docker.

In the following sections, we will introduce each of these offerings to provide you with an overview of their strengths and ideal usages. Each of the offerings provides you with the same SQL Server relational engine internals for storing and querying data using T-SQL. Each will have the same **data manipulation language (DML)**, with only minor differences in **data definition language (DDL)** due to physical implementation differences. While some of the management tools and methods supported by each platform are different, the primary difference is based on how **managed** the service is.

SQL Server on Azure Virtual Machines

SQL Server on Azure Virtual Machines[7] (or Azure SQL VMs for short) and indeed any of the Azure Virtual Machines offerings are considered IaaS. This is because Microsoft manages the hardware infrastructure, but you manage the software. As the DBA managing the server, it is generally no different than managing SQL Server on a computer that resides in your own server room.

When you create an Azure SQL VM, you are given the opportunity to use a pre-built VM image that has a supported version of SQL Server pre-installed, or you can choose to bring your own media to install from. There are licensing differences and benefits to both models, but we will not even begin to try to cover licensing in this book. Here are some more insights offered by Microsoft on Azure VM licensing[8].

Whether you use a pre-built image or bring your own software, the VM can take advantage of some automation by using the SQL Server IaaS Agent Extension (not currently available on an Azure VM running Linux at the time of this book's publishing), which provides automated backup and patching capabilities, as well as configuration assistance with Azure Key Vault integration to store encryption keys outside of SQL Server. You are also fully able to use any method you wish for these tasks, including SQL Server Maintenance Plans or even third-party backup scripts and tools. Some additional tooling may be necessary in any case, because backups are just part of the regular upkeep needed for a healthy database that is even lightly used. The Agent Extension (along with several other features) is enabled automatically when using a pre-built image, or by registering[9] your VM with the SQL VM resource provider.

It is important to catch the distinction between automation and a managed service. Automation provides tools that you can use to make managing your server easier. With the PaaS model of the next two Azure offerings, you don't need to monitor to see whether backups have failed, nor do you even need to do anything to ensure that your server is backed up. The Azure platform management system backs up your server based on the settings you choose (and you can't even accidentally choose not to back up at all either). With IaaS, only the hardware is truly managed by Microsoft. It is your responsibility to back up your databases and make sure those backups can be restored, even when using the SQL Server IaaS Agent Extension.

> **Note**
>
> Managing and supporting are two different concepts. A supported service means the host will help you if the software is not working properly. A managed service will have the host in charge of making sure things work properly based on your configuration.

An SQL Server VM gives you a highly compatible method to lift and shift many workload types to the cloud. This includes transactional workloads capturing customer orders or business intelligence workloads using analytics features such as Machine Learning Services, Reporting Services, Analysis Services, and so on. This is because the Azure VM presents itself as very much the same as your on-premises hardware, the only real difference being how you configure networking and security over the internet to work with your local security infrastructure. For SQL servers that use SQL authentication, the application will require little, if any, change, but using Active Directory will require some configuration. (Using SQL-based authentication is not considered as good as using Active Directory integration for many reasons. Chapter 3 will cover security, including integrating with your existing Active Directory.)

One major choice you have, beginning with SQL Server 2017 and continuing into SQL Server 2019, is which OS to choose. Beyond Windows Server, SQL Server will run on **Red Hat Enterprise Linux (RHEL)**, **SUSE Linux Enterprise Server (SLES)**, and **Ubuntu**. This allows SQL Server to be run on any OS used by an organization that's heavily invested in **open-source software (OSS)**, while still providing virtually all the database features of the Windows version.

Before we dig deeper into the topic surrounding OS choice for virtual machines, there are still two configurations of SQL in Azure that we want to cover, because they offer specific benefits that the IaaS platform may not.

Azure SQL Managed Instance

The Azure SQL Managed Instance deployment option is a PaaS database offering that sits right in the middle of the Azure SQL offerings in terms of management and is one of two Azure SQL PaaS offerings. It targets scenarios where the customer needs much of the rich functionality of the full SQL Server product but desires the value of the platform-managed services model as well.

One of the biggest differences between the managed instance model and the Azure SQL VM model is that a managed instance will always be on the most up-to-date version of SQL Server (a database can be set to an earlier compatibility level if needed). This means, at the time of this book's writing in early 2020, managed instance–hosted databases provide the user with at least everything that SQL Server 2019 has to offer, up to the latest public **cumulative update (CU)**, including all the new and improved features in SQL Server 2019 (we will provide an overview later in this chapter). Soon after the next version of SQL Server is released, managed instances will be upgraded as part of the offering.

While it is extremely rare that Microsoft removes features from SQL Server (in fact, backward compatibility is an important part of their story), there are frequent changes in how the engine optimizes queries, so be sure to keep up with performance changes in your application. This process of determining and mitigating performance issues after releases is made all the easier by the Query Store feature[10].

The managed instance model will automatically maintain backups[11] of your data, restorable to a point in time, based on the level of retention you configure. It safeguards your backups from disaster by using **read-access geo-redundant storage (RA-GRS)**.

While backups are part of the managed instance package, geo-replication (that is, maintaining active copies of your data in multiple regions in case of catastrophic failure) is not. The managed instance option provides you with Auto-Failover Groups[12] you can configure like in SQL Server 2019. This lets you configure your server to fail over to a different Azure datacenter if desired.

In many ways, a managed instance looks and behaves just like a typical SQL Server, particularly when dealing with DML code, instance security, and tools such as SQL Agent. (SQL Agent is limited in the types of jobs it can schedule, due to a lack of OS and file system access.) There are important considerations, such as the inability to restore a database from a managed instance to a local or VM SQL Server instance or to use **SQL Server Integration Services (SSIS)** in the same way as you do with a typical instance. (SSIS projects will work using Azure Data Factory[13], but the infrastructure of the SSIS DB does not exist natively in the managed instance itself. For more details, you should refer to Microsoft's guide to migrating SSIS packages to Azure SQL Managed Instance[14]. Any code that needs access to the file system, such as for processing files for import or export, will need to be modified to work with Azure Blob storage[15] instead.

Managed instances provide a lot of parity with an up-to-date version of SQL Server, providing a way to lift and shift many workloads to the cloud with a reasonable amount of work. However, there are limitations, such as always being on the most recent version of SQL Server, a lack of access to the file system, and not being able to restore to a local SQL Server instance that may make this untenable for many scenarios, leading to using the IaaS VM offering.

In the next section, we will look at one more Azure SQL offering that goes deeper into the managed aspect of PaaS offerings, even beyond what you have in a managed instance. It may require more adjustment from an existing on-premises environment but offers conveniences in management that leave you to mostly design and develop tables and code.

Azure SQL Database

Azure SQL Database[16] (or SQL Database for short) is the fully managed version of SQL Server, falling deeper into the category of PaaS than the managed instance model. Using SQL Database, you are provided with a database container (or containers, in more complex configurations) in which to create tables and coded objects. The T-SQL features of SQL Database may actually be more advanced than what you get with the other offerings, as they often release features to SQL Database first.

There are many configurations to choose from in terms of performance and size, starting small and scaling up to very large database sizes. In fact, using Hyperscale[17] for SQL Database, this platform's offerings can currently support up to 100 TB of data, so the amount of space and computing power is not generally a limiting factor.

The SQL Database offering is tailored to cloud applications with the least amount of legacy management dependencies. This does not mean that your configuration must be one simple database, however, and at the time of writing this book, there are three options for how you can set up your SQL Database environment:

- **Single database:** A single database that can be used to store data. This will feel like an SQL Server instance to the user/administrator, with access to a TempDB and master database. If you have connected to a contained database in SQL Server, it is conceptually similar except the boundaries are far more firm. Compute resources are available via pre-provisioned or serverless options[18], ensuring sufficient resources for both consistent and highly unpredictable workloads.

- **Elastic Pool**: Multiple independent single-database configurations that share the same set of computing resources. This allows you to have low-use and high-use pattern databases in the same pool, where both databases can use up to the maximum resources when needed, but not pay for large amounts of resources dedicated to one database when it is only rarely needed. This is particularly effective when the databases see heavy utilization at different times of the day.

- **Database server**: A group of single databases and elastic pools, banded together for administrative purposes, for things such as networking, security principals (logins), policies, and so on.

As the consumer, you are getting the SQL Server database as a platform to store data.

Some aspects of the table structures you create on Azure SQL Database may need to be different from the other offerings. One key difference is that every table will require a clustered index to enable the use of replication of your data to redundant storage. (Using clustered indexes on all tables is considered a best practice in most cases anyhow.) You have no control over where your data is located (outside of the region of the world), or the computer the data is located on. This is all done for you.

Just like managed instances, backups are part of the managed package, but additionally, high availability via geo-replication is also supported as part of the management of your database, with very little configuration. Microsoft's guide offers more insights on active geo-replication[19].

A feature that is specific to SQL Database that is particularly useful is automated indexing[20]. Using the automatic index management feature, SQL Database can apply a **CREATE** or **DROP INDEX** automatically based on what the optimizer has recognized as a given need, then monitor that change to see whether it has helped or harmed performance and adjust accordingly.

The SQL Database offering provides the most management for you but may not exactly match the needs of an organization with a well-established working system in an SQL Server instance.

Now that we have covered these three offerings, in the next section, we will outline some of the important differences between them.

SQL Server in Azure comparisons

All three of the Azure SQL offerings that we have introduced in this section are based on the same SQL Server 2019 database engine, but there are key differences between them. In the following table, we list some of the key differences to remember when considering which option to choose:

	SQL Server on Azure VMs	Azure SQL Managed Instance	Azure SQL Database
Model	IaaS	PaaS	PaaS
Core Usage	Moving workloads to Azure with the least amount of change	Migrating to an environment that needs far less management than IaaS (but still retains the general SQL Server configuration)	New applications, often cloud-based, that have no legacy dependencies on SQL Server instance constructs
Backups	Managed by DBA	Automatic (can make manual copy-only backups to Azure Blob storage)	Automatic
File-Level Access	Full	Limited to accessing Azure Blob storage	N/A
OS	Microsoft Windows, Linux (RHEL Server, SLES, and Ubuntu)	N/A	N/A
Task Scheduling	SQL Agent	SQL Agent (Limited types of jobs supported)	Azure Automation
SQL Server Engine Version	Supported release (or bring your own media to install any version)	Greenfield: Latest release on the latest CU, databases can be used past compatibility level	Greenfield: Latest release, latest CU, compatibility with current release only
High Availability	Manually configured	Built-in (Disaster recovery by using Always On Availability Groups)	Built-in

Figure 1.1: Azure SQL offerings

In the next section, we will highlight the key new features that were included in SQL Server 2019 that make it a worthwhile upgrade from earlier versions (including from the most recent version before 2019: SQL Server 2017).

SQL Server 2019 highlights

SQL Server 2019 has a lot of new and enhanced features that make it not only a world-class relational database engine, but a world-class data platform. In this section, we will be taking a brief look at what SQL Server 2019 adds on from previous versions, as well as the features that have been added to running SQL Server on a Linux-based platform in SQL Server 2019.

You can read more about the new features available in SQL Server 2019 on Microsoft documentation[21] or the Microsoft SQL Server 2019 Technical white paper[22].

Intelligence over all of your data

Intelligence over all of your data is a phrase that you can find on multiple Microsoft websites describing an important SQL Server feature for integrating your disparate data sources: PolyBase. PolyBase[23] is a feature that can be used to *virtualize* data from external sources including Azure Cosmos DB, Azure Blob Storage, and starting in SQL Server 2019, data in SQL Server, Oracle, Teradata, and MongoDB. Once connected and configured, you can query this data just like any normal relational table (as well as joined with your local relational data).

Additionally, SQL Server 2019 introduces Big Data Clusters[24], providing scale-out capabilities with clusters of SQL Server, Apache Spark, and **Hadoop Distributed File System (HDFS)** data, allowing reading and writing of large quantities of data stored in SQL Server or big data sources.

Both features allow you to bring relational data and big data together to provide a uniform platform for data processing. This allows you to use tools such as Machine Learning Language Extensions, AI with Machine Learning Services, Reporting Services, and even SQL Server's primary language T-SQL with data from very differently formatted data structures, without doing any copying or transformation of data.

Enhancements in developer experience

SQL Server 2019 includes several important improvements to the developer experience that help to make developing data-based solutions easier:

- **UTF-8 support**: Allows you to store Unicode data in the **char** and **varchar** data types via new collation support. This allows you to store data in the very popular Unicode encoding format natively, not requiring translation into the UTF-16 standard that is used in **nchar** and **nvarchar** columns. For more information, refer to the Microsoft documentation[25] that covers collation and Unicode support in SQL Server. For an in-depth commentary on UTF-8 in SQL Server, refer to the enlightening blog[26] by Pedro Lopez.

- **Machine Learning Services**: Machine Learning Services[27] enables R and Python support, allowing T-SQL to employ machine learning models where your data lives. It allows processing data at the partition level, rather than only at the object level, enabling parallel processing. Additionally, Machine Learning Services can now be used with Failover Clusters.

- **SQL Server Language Extensions and the Extensibility Framework**: Improvements made to allow additional languages to be run in a very similar manner to R and Python, which are used by Machine Learning Services. This gives developers more choices for running established code right in the SQL Server engine. Currently, only Java is supported, but more languages will follow. For more details, check out Microsoft's overview of language extensions[28].

- **Lightweight query profiling**: Improvements to live query plan gathering[29] to make it cheaper to get statistics and progress on currently executing SQL statements.

- **SQL Graph enhancements**: SQL Server graph database capabilities[30] allow for the creation of nodes and edges (many-to-many relationships), which are often needed for applications where a more traditional relational schema is too rigid and complex to query. Improvements now allow edge constraints (foreign keys on edges) and query improvements to query for nodes that are multiple edges away from one another.

Performance enhancements

SQL Server has been a leader in performance for years, with great backward compatibility at the code level, and tremendous leaps in performance in every new version. SQL Server 2019 brings with it even more improvements to performance, many built on enhancements from recent releases. For example, memory-optimized tables can bring tremendous performance improvements when storing data, and in SQL Server 2019 that feature has been leveraged to enhance the metadata for TempDB.

In this section, we will briefly look at some of the performance enhancements in SQL Server 2019:

- **Enhancements to Intelligent Query Processing (IQP)**: IQP[31] is a family of loosely connected technologies designed to improve your SQL Server, many without any changes to your code. Several new methods of improving performance, such as expanding batch mode to include rowstore structures and inlining scalar functions, have been added.

- **Accelerated Database Recovery (ADR)**: ADR[32] dramatically reduces the time required to return control to the user in a rollback of a large data change/recovery on restart, doing much of the work asynchronously.

- **Hybrid Buffer Pool**: Hybrid Buffer Pool[33] provides support for **persistent memory modules (PMEMs)**, which allows the engine to use PMEMs in the classic roles typically played by RAM and the data file, eliminating the need to checkpoint data from memory, greatly enhancing performance for scenarios where very high performance is required.

- **Memory-Optimized TempDB Metadata**: The metadata for the TempDB[34] database can be altered to use in-memory data structures, removing bottlenecks that occur when rapidly and concurrently creating a large number of temporary tables.

Security improvements

Security is (or at least should be) one of the most important concerns for any data engineer/administrator. SQL Server 2019 introduces several important security improvements, building on improvements in recent versions of SQL Server, such as Always Encrypted[35], row-level security[36], dynamic data masking[37], transparent data encryption[38], and far more[39] than we will cover in this book.

Changes in SQL Server 2019 to security include:

- **Always Encrypted with secure enclaves**: An enhanced form of Always Encrypted[40] that allows computations/searches to occur on the server side on encrypted string data, but still never shares the plaintext with the user or administrator without the required key.

- **Data classification and auditing**: This starts with an **SQL Server Management Studio (SSMS)** tool to help locate, classify, and tag sensitive data that may need to be handled specially. Next, to know when sensitive data is being accessed, SQL Server Audit[41] includes a new field in its output that indicates the data sensitivity of data that is included in the audit output. For more details, refer to this Microsoft guide[42] on data discovery and classification.

- **Simplified certificate management**: Certificate management[43] is integrated with SQL Server Configuration Manager.

High Availability/Disaster Recovery (HADR)

As your demand for around-the-clock access to data increases, the following features are designed to make usage of SQL Server even more possible during maintenance and during a failover to a different server:

- **Index maintenance enhancements**: SQL Server 2019 adds to the types of indexes that can be rebuilt online to include clustered columnstore indexes, as well as allowing you to pause and resume rowstore index rebuilds. For more details, check the information[44] provided by Microsoft.

- **Availability group enhancements**: Availability groups[45] are an HADR feature that was first introduced in SQL Server 2012. They allow you to maintain copies of your database in a different location to fail over to when there is a failure in your primary database/server (as well as other uses). SQL Server 2019 increases the number of synchronous secondary replicas from three to five. Automatic client redirection[46] has been added, so clients can fail over without changing the connection string. Additionally, there have been licensing improvements[47] for Software Assurance customers pertaining to HADR scenarios.

Platform of choice

SQL Server has been around for over 20 years on Windows, but in SQL Server 2017, the platform choices grew to include Linux. SQL Server 2017 on Linux included most of the primary features of a relational engine that customers needed, but not all of them. SQL Server 2019 adds most of the features that were missing in SQL Server 2017.

Features added for SQL Server 2019 on Linux include:

- **Replication**: Data is allowed to be copied automatically between databases on the same or different instance (including Windows instances). Transaction and snapshot replication is now in Linux.

- **Distributed Transactions**: Enables transactions that extend beyond the confines of the instance.

- **Change Data Capture**: Maintains a history of changes to data in a database, commonly for processes duplicating data where replication doesn't make sense.

- **Extended Active Directory Support**: Adds support for third-party Active Directory integrations.

- **Machine Learning/Language Extensions**: Adds the ability to run R, Python, and Java inside the SQL Server engine.

- **PolyBase**: Ability to query external data and leverage data virtualization using T-SQL, as described earlier.

For a complete list of improvements to the Linux version of SQL Server with links to more details, check out the information at the Microsoft documentation[48].

Beyond Linux on a VM or on-premises server, improvements have been made when installing SQL Server on a container. Azure Container Registry[49] provides a location to manage containers for Docker and Open Container Initiative images.

SQL Server is also now available[50] on Red Hat Enterprise Linux 8, as well as using Red Hat Universal Image Containers.

Finally, when using containers, SQL Server 2019 does not need to be started as a root container[51] in Linux, providing a more secure experience.

SQL Server IaaS scenarios and use cases

With all the choices for how to deploy and implement SQL Server for your organization, both on-premises as well as through the Azure SQL offerings, why choose SQL Server on an Azure VM? While the more managed versions of SQL Server provide a lot of benefits, they have downsides if you are heavily invested in SQL Server on-premises because they can require you to change your infrastructure considerably. (Even the managed instance offering may be too restrictive for many organization's needs.) The IaaS model allows you to use the entire SQL Server 2019 feature set in a way that will work mostly as your on-premises model has for years.

The IaaS model is natural to existing DBAs while reducing (or even eliminating) the need to house and manage server hardware in your organization's premises. There is also licensing value as well, because SQL Server VMs can be adjusted in power and storage relatively easily, as well as licensed to pay as you go, allowing you to start and stop as you desire and incurring far smaller costs when VMs are not running.

In this section, we will look at three use cases for SQL Server 2019 in an Azure VM that will immediately benefit your organization, with the least friction with what your staff already know:

- **Lift and Shift**: Keeping your applications pretty much as is, moving to Azure.

- **Extending On-premises Environment to the Cloud**: When you need to add a server to your environment and do not wish to add more hardware to your estate.

- **Development and Test Environments**: Developing new code and testing it, perhaps in the latest SQL Server version or a different edition (for example, if you are on standard edition and want to see what the effect of using Enterprise Edition would be).

These scenarios are ones that you are likely to use IaaS for. Not for radical change to what you currently do, but rather to make use of the knowledge and skill set of existing database human resources, all while reducing the need to manage and maintain hardware on your premises, either for long-term or short-term utilization.

Lift and Shift

For many organizations, keeping hardware up to date is a daunting task. You purchase a server one year, and it becomes outdated very quickly. Three or four years later, when the hardware is very outdated, you finally have the budget to upgrade the hardware. Then, you can perform the upgrade, including creating servers, moving data around, and so on. All of this is a time-consuming, costly process, even if you are already virtualizing your server resources on-premises. At the same time, moving your environment to a fully managed service, even a managed server, can be complex or even not possible, depending on your requirements. (For example, if you need to use the file system or other executables on the same server, it is not possible in a managed server or SQL Database.)

In a lift-and-shift operation, you get one or more servers that you can configure exactly as your on-premises server were configured. The same drives, directories, services, and so on, even including third-party executables.

What you don't have to worry about is the scalability of your server or how to keep up with the latest hardware. An Azure SQL VM can be sized as needed, to any level of hardware you need, and resized[52] if needs change.

Now the burden of managing the hardware lies with the Microsoft Azure management team, leaving your organization with more time to build and test quality software.

Extending your on-premises environment to the cloud

If moving your entire datacenter to Azure is not something your organization is ready for, that doesn't mean that SQL Server on an Azure VM is not useful. There are a few ways you might extend your environment using Azure VMs.

A primary way would be as a disaster recovery site. You can have a reasonably low power server that you copy your data to, perhaps utilizing Always On availability groups[53] using a hybrid on-premises and cloud configuration. If the server needs to recover from a disaster, you ramp up the resources, point clients to the Azure VM instead of the local server, and you are ready to go.

Another common use is when you need an extra server to meet a need (possibly even a short-term need), but do not have the inclination to procure and configure a server and permanent SQL Server licenses. Once you have the VPN gateway[54] configured, adding a VM-based server in Azure is very much like adding one locally. The main difference is that Microsoft does the heavy lifting of setup for you; you just determine how much power you need (and adjust as necessary). And since SQL Server licensing can be built into charges, if the need is temporary, you do not need to buy a full-price license for SQL Server. Getting rid of a server you no longer need is now simply a matter of clicking delete on the portal (and verifying that you do actually want to delete the server, naturally).

Development and test environments

The last use case we will present is about using Azure SQL VMs for developing and testing new software. Creating a new SQL Server VM can be done in a very short amount of time, and subsequent spin-up time can be further shortened by using templates to configure everything as you need it.

Developers can get a new VM with an SQL Server Developer Edition license, then quickly load executables and test data on it. Once done with testing, they can shut down the server and only pay for storage costs. If you need to start over from scratch, it is easy to delete the server and start the process over with a fresh install.

In the next section, we will look at how we can choose the best OS for your SQL Server VM.

Choosing an OS for SQL Server in Azure VMs

When you create a new SQL Server on an Azure VM, among the first choices you have is which OS to choose. In this section, we will look at some of the thought processes to go through when deciding whether to choose Windows Server or one of the Linux distributions (also referred to as distros) for your SQL Server instance.

SQL Server can run on Windows or Linux, and what might be surprising is that it is the exact same SQL Server engine code base. Of course, Windows and Linux cannot run the same binaries natively, so Microsoft built the **SQL Platform Abstraction Layer** (**SQLPAL**), which makes this a reality. For more details on SQLPAL and how it is used, the blog[55] by the SQL Team offers interesting insights.

While SQL Server on Windows and Linux have the same code base, there are differences to be understood. In this section, we will look at the reasons for choosing either Windows or Linux, and then discuss the differences between the two implementations.

Reasons to choose either Windows or Linux for SQL Server

This book's focus is on using Microsoft SQL Server on an IaaS platform, and the fact that SQL Server runs on either OS may sound fishy to you. You may be suspicious that they want to get you on Linux and then lure you to Windows over time. It is important to realize that this is not the case: they are both the same SQL Server product and are equally supported.

The primary deciding factor on which OS to use is actually based on the comfort level of the company that will be using it. While there are a few differences in feature sets, there is no compromise in quality or performance with either choice. Much like the decision process in choosing between the IaaS model or the PaaS model, the choice to use Linux for the server OS is more a matter of your organization's needs, rather than being based upon any objective benefit.

If an organization is not comfortable with Windows, it is going to be harder to implement SQL Server on Windows, even using a pre-built image. If you are comfortable with Linux, the process to install and configure SQL Server is more like what you are used to when installing Linux software. As we will see in the following section, while the functionality to the user is going to be almost exactly the same on either OS, the differences in how you configure SQL Server are significant.

One thing to note is that **Open-source Software (OSS)** developers often prefer Linux because they can install and run additional OSS packages in the same environment with their database server as part of a solution. These packages may or may not run on the Windows platform. SQL Server on Linux lets them reap the benefits of SQL Server's mature data storage engine to build applications that run faster, are more secure, have tight integration with machine learning, and so on.

> **Note**
>
> The goal of SQL Server on Linux is not to kill the Windows version, but to broaden the audience, making the product speak the implementation language of a new audience while providing the same SQL Server T-SQL language to all.

The only reason that you may not be able to feasibly choose Linux for your OS is that some features are not yet supported (for example, Merge Replication, or having multiple instances on a single server), some that will likely never be (such as **FILESTREAM** and **FileTable**, both of which interface tightly with the Windows OS), and others that will require new binaries to be created (or another abstraction layer), such as Analysis Services. Any need for these features, even something like Reporting Services, which runs as a separate executable, would currently require an entire extra license for SQL Server to run on a separate Windows Server, which could be cost-prohibitive.

The final issue to bear in mind when choosing an OS for SQL Server is cost. Cost is a very complicated discussion because costs come from many different places, some more obvious than others. There are obvious cost differences you can compare empirically, such as the hourly rate of running a VM on Windows versus one with Linux. However, if you must hire new staff and train them on how Windows works, or pay consultants every time something isn't working, the costs may be prohibitive. The same can be said about Linux.

In the end, the most compelling reasons to choose between Linux and Windows are to do with your comfort level with each OS and whether you need specific features that may not be available on Linux (a list that we noted earlier shrank considerably with SQL Server 2019.)

Differences between SQL Server on Linux and Windows

Beyond the obvious deep differences in how the different OSes behave (even the different versions of Linux have their own ways of doing things), there are some differences between the Windows and Linux versions of the product.

In this section, we are going to highlight some of the differences between SQL Server on Linux and Windows, whether you use a pre-built SQL Server VM image from Azure or install SQL Server on your own on-premises computers. The following table contains a list of key differences between SQL Server on Windows and Linux:

Windows	Linux
Installation/configuration typically done using the **graphical user interface (GUI)** by most DBAs.	Installation/configuration always done using the **command-line interface (CLI)**.
Installation is done using one executable.	Features are installed using multiple package managers. For example, the database engine, full text search, Integration Services, SQL Server Agent, PolyBase, and so on are all in individual packages.
Can run all SQL Server services natively.	Several external SQL Server services are not currently supported (Analysis Services, Reporting Services, and so on).
Multiple instances of SQL Server on one Windows server.	A single instance per Linux server.
Code executed in the engine can access the OS (XP_CMDSHELL, CLR assemblies, FILESTREAM/Filetable).	**Features that access the file system (other** than backup/restore) are not enabled.
SQL Server Agent can send alerts when issues occur or certain errors are raised.	SQL Server Agent cannot send alerts.
SQL Server Agent can execute command files, **PowerShell scripts, SSIS packages,** SQL Server Analysis service (SSAS), or SQL Server Reporting service (SSRS) actions natively.	SQL Server Agent cannot execute command files, **PowerShell scripts, launch SSIS** packages, or perform SSAS or SSRS actions.
Databases can use Stretch DB to store infrequently used data in Azure Storage, instead of using local storage.	Stretch Database is not supported.

Figure 1.2: Differences between SQL Server on Linux and Windows

A few of these differences warrant a little bit of discussion; most importantly, we'll consider the way you install and configure SQL Server and some feature-set limitations of SQL Server on Linux.

Installation/configuration

While the experience of the typical user employing T-SQL or using an application will be very nearly 100% the same, there are some major differences between SQL Server 2019 on Linux compared to running it on Windows. Obviously, the biggest difference is that in Windows, most SQL Server DBAs will be used to working with a GUI rather than the CLI. Hence, changing from Windows to Linux can be very a large paradigm shift, particularly during installation but also even when choosing where to locate database files. However, if you are used to Linux, the method of server and instance installation/configuration should generally be obvious to you; the same goes for how the file system works.

While it is true that there is a version of Windows that you manage mainly from the command line (Windows Core) that SQL Server can execute on, it is not typically used because Windows administrators are generally used to managing the server via the GUI (in the same manner as their Windows and even Macintosh computers that they regularly use). Even then, however, the commands to install on either platform differ in that the Core SQL Server installation is one executable with many parameters, rather than requiring multiple commands to install different features.

When you install the Linux image with SQL Server the first time, it will be necessary to access the server via what is basically a command terminal. Features are added and configured individually using command-line tools instead of **setup.exe**. Once you have the server installed, you can access the SQL Server instance on the Linux computer using SSMS, Azure Data Studio, SQL Server Configuration Manager, and the other GUI tools on your Windows computer. If you do wish to run tools on a Linux GUI, SSMS will not work, but Azure Data Studio[56] will.

Feature set

Probably the most compelling reason for choosing Windows over Linux is if you need some of the external services that are not on Linux. While the engine is the same, there are several services that have their own binaries and are not a part of the core SQL Server engine. For SQL Server 2019, this list includes:

- Reporting Services
- Analysis Services
- Data Quality Services
- Master Data Services

You would need another SQL Server license to run these services on a different computer, which might increase your costs greatly. None of these services are needed for a typical OLTP database, but if you are looking to implement a **business intelligence (BI)** solution together with SQL Server on Linux in Azure VMs, cloud-based options to augment your VM include Power BI and Azure Analysis Services.

Summary

In this chapter, we looked at SQL Server 2019 and the various Azure SQL offerings that are available to deploy your data. There are multiple models available, from a very managed PaaS service such as Azure SQL Database, to the manual-lite managed instance, to the manual and customizable IaaS-based Azure SQL VM. The rest of this book will focus on SQL Server on an Azure VM, whether on Windows or Linux.

SQL Server 2019 on Linux provides customers with a choice of platform, allowing them to tailor SQL Server to the needs of their personnel where much of their software is run on the Linux open-source OS. SQL Server is not itself open source, but Microsoft is embracing the OSS community like never before.

In the next chapter, we will look at the different options for deploying a VM on Azure for SQL Server. Choices such as series of VM, storage types, and installation will be covered, taking into consideration the different workload types that may need to be supported.

Chapter links

1. https://bit.ly/2XiCe5Y
2. https://bit.ly/3g8G8Hh
3. https://bit.ly/2TqJ4Fl
4. https://bit.ly/2ZmMaOw
5. https://bit.ly/36gBLFq
6. https://bit.ly/36ifoiT
7. https://bit.ly/36pvzLC
8. https://bit.ly/3e4YxTp
9. https://bit.ly/2Xc8GHc
10. https://bit.ly/3ga5Yuw
11. https://bit.ly/3cQoHsN
12. https://bit.ly/3cOeUUo
13. https://bit.ly/3e6t7wc
14. https://bit.ly/2ZnmEIZ
15. https://bit.ly/2LIiBPz
16. https://bit.ly/2XeRzVk
17. https://bit.ly/2yjMHWk
18. https://bit.ly/2LMUU8A
19. https://bit.ly/2LM0zMe
20. https://bit.ly/2Zpej7M
21. https://bit.ly/3ehBFjZ
22. https://bit.ly/2WRIjaK
23. https://bit.ly/36oCaFZ
24. https://bit.ly/2XcaHTM
25. https://bit.ly/3d29K72

26. https://bit.ly/2ZoWC88
27. https://bit.ly/2TsFjzr
28. https://bit.ly/2XfwSbH
29. https://bit.ly/36mhq1M
30. https://bit.ly/3bMWWjD
31. https://bit.ly/2LI5UnI
32. https://bit.ly/2WPvu0D
33. https://bit.ly/2LNzFn2
34. https://bit.ly/3bJCiRy
35. https://bit.ly/2zicH4P
36. https://bit.ly/36gHmvw
37. https://bit.ly/2z7DNMf
38. https://bit.ly/2AFx6kH
39. https://bit.ly/3e9YIgt
40. https://bit.ly/36mikLI
41. https://bit.ly/2Zoid0H
42. https://bit.ly/3gcfwVJ
43. https://bit.ly/2AO4QfW
44. https://bit.ly/2ZrNIqv
45. https://bit.ly/2LLyTHl
46. https://bit.ly/2LNBala
47. https://bit.ly/2XgnoNp
48. https://bit.ly/2AOPp7d
49. https://bit.ly/2XdyvXm
50. https://bit.ly/2TqTwwK

51. https://bit.ly/3bQMozS

52. https://bit.ly/2XdyOBu

53. https://bit.ly/3cR75NG

54. https://bit.ly/36iLjQk

55. https://bit.ly/2Tkvdk3

56. https://bit.ly/2LLhGh5

2

Getting started with SQL Server on Azure Virtual Machines

By Allan Hirt

Virtual machines (VMs) are the core of all **infrastructure-as-a-service (IaaS)** deployments. A VM in Azure is similar to a VM premises using a hypervisor. The main difference is that Microsoft maintains the hypervisor and its related infrastructure in Azure. A VM also means that whether you choose to have Microsoft manage selected administration tasks, such as backups, or you perform those tasks yourself, it is still an operating system (Windows Server or a distribution of Linux) and installation of SQL Server that needs to be administered, maintained, patched, made available, and so on.

This chapter will help you understand why you should consider IaaS and covers how to approach choosing the right virtual hardware for a VM with SQL Server in Azure.

The benefits of deploying SQL Server using IaaS

Besides IaaS, there is another deployment method for SQL Server: platform-as-a-service (PaaS). Azure SQL Database, or Azure SQL Managed Instance, is PaaS. With PaaS, there is no operating system (OS) or SQL Server instance you need to manage, which includes things such as patching. All of that is done for you. Where that does not work for some is that you may need control over the OS and/or SQL Server choices (version or edition) due to standards, licensing, or other requirements. PaaS provides a more packaged solution that fits the needs of many without needing a dedicated OS and SQL Server instance.

Since IaaS is just a VM with an OS, you can deploy whatever supported combination of OS and SQL Server you desire. One principal difference between IaaS and PaaS is that any automatic administration you would want done by Microsoft is opt-in, whereas PaaS is done for you and you have limited configuration choices. One important reason why many choose IaaS is the ability to make choices you could not otherwise make with something like Azure SQL Database.

That choice extends to being able to use other Azure services, such as Azure Backup[1], Azure Security Center[2], Advanced Data Security for SQL on Azure VM[3], including vulnerability assessment and Advanced Threat Protection, Azure Site Recovery[4], and the SQL Server on Azure Virtual Machines resource provider (covered in a later section). This means that you can take advantage of the best of what Azure has to offer for IaaS-based SQL Server deployments.

IaaS facilitates installations of SQL Server in cloud-first environments, "lift and shift" scenarios where you have a requirement to deploy the same version/architecture as on-premises but are migrating to Azure, and, more importantly, hybrid scenarios where IaaS VMs become an extension of on-premises.

For example, you may still be primarily on-premises for most SQL Server installations right now. However, you want to start moving toward using Azure and also have a need for more robust disaster recovery. If you currently use Always On availability groups, one or more IaaS-based replicas could be added to Azure (along with any other required infrastructure, such as Active Directory Domain Services) to extend the existing architecture up to the cloud. Refer to this Microsoft documentation[5] to read more about the new high availability and disaster recovery benefits for SQL Server.

All standard SQL Server deployment scenarios apply in IaaS as they do on-premises if you are implementing physical servers or VMs. Most considerations for deploying SQL Server properly on-premises are the same, with slight variations due to Azure, some of which are documented in this chapter and others throughout this book. The rest of this chapter will focus on how to choose a VM and approach things such as sizing.

There is one other benefit associated with choosing Azure for your IaaS SQL Server platform: Microsoft will continue to provide Windows Server 2008 and 2008 R2 as well as SQL Server 2008 and 2008 R2 extended security updates if you still require those versions and need to migrate those workloads to VMs in Azure.

Deployment choices for IaaS

This section will cover the different ways in which you can deploy a VM in Azure, along with considerations and best practices that are applicable no matter which method is used, as well as briefly talk about licensing SQL Server in Azure.

Deployment methods

There are three options for deploying a VM in Azure for use with SQL Server:

- Choose a pre-built template, also known as an image, with SQL Server already installed.

- Choose a template with the OS but without SQL Server installed, which will be done after the VM is built.

- Build a custom image using a virtual hard drive and upload it to Azure.

No matter which of the three deployment methods you choose, you can deploy using Cloud Shell (Bash or PowerShell)[6], also known as the Azure command-line interface (CLI), Azure portal[7], Azure Resource Manager templates[8] that are written in JSON, and the Azure PowerShell module[9]. Builds can be automated and incorporated into DevOps processes, which is often a best practice in many environments.

The remainder of this section will describe each of these methods in further detail.

Using an image with SQL Server installed

Microsoft provides images in Azure that have SQL Server already installed. Using one may save time and effort and provide a level of confidence in the overall configuration. The OS, versions, and editions of SQL Server available can be seen by querying Azure or looking in Azure Marketplace using the Azure portal. One example query using the Azure CLI that shows all Ubuntu images in the East US 2 region with SQL Server Enterprise Edition, as of the time of writing of this chapter, is as follows:

```
az vm image list --location eastus2 --offer Ubuntu --publisher SQLServer
--sku enterprise --all --output table
```

The results are shown in *Figure 2.1*. A similar query could be executed for Windows Server or any other distribution of Linux or edition of SQL Server:

```
allan@Azure:~$ az vm image list --location eastus2 --offer Ubuntu --all --output table --publisher SQLServer --sku enterprise
Offer              Publisher           Sku         Urn                                                                Version
-----------------  ------------------  ----------  ----------------------------------------------------------------  ----------
SQL2017-Ubuntu1604 MicrosoftSQLServer  Enterprise  MicrosoftSQLServer:SQL2017-Ubuntu1604:Enterprise:14.1.190409  14.1.190409
SQL2017-Ubuntu1604 MicrosoftSQLServer  Enterprise  MicrosoftSQLServer:SQL2017-Ubuntu1604:Enterprise:14.1.190611  14.1.190611
SQL2017-Ubuntu1604 MicrosoftSQLServer  Enterprise  MicrosoftSQLServer:SQL2017-Ubuntu1604:Enterprise:14.1.190709  14.1.190709
SQL2017-Ubuntu1604 MicrosoftSQLServer  Enterprise  MicrosoftSQLServer:SQL2017-Ubuntu1604:Enterprise:14.1.190813  14.1.190813
SQL2017-Ubuntu1604 MicrosoftSQLServer  Enterprise  MicrosoftSQLServer:SQL2017-Ubuntu1604:Enterprise:14.1.190910  14.1.190910
SQL2017-Ubuntu1604 MicrosoftSQLServer  Enterprise  MicrosoftSQLServer:SQL2017-Ubuntu1604:Enterprise:14.1.191008  14.1.191008
SQL2017-Ubuntu1604 MicrosoftSQLServer  Enterprise  MicrosoftSQLServer:SQL2017-Ubuntu1604:Enterprise:14.1.191210  14.1.191210
SQL2017-Ubuntu1604 MicrosoftSQLServer  Enterprise  MicrosoftSQLServer:SQL2017-Ubuntu1604:Enterprise:14.1.200114  14.1.200114
SQL2019-Ubuntu1604 MicrosoftSQLServer  enterprise  MicrosoftSQLServer:SQL2019-Ubuntu1604:enterprise:15.0.191104  15.0.191104
SQL2019-Ubuntu1604 MicrosoftSQLServer  enterprise  MicrosoftSQLServer:SQL2019-Ubuntu1604:enterprise:15.0.191210  15.0.191210
SQL2019-Ubuntu1604 MicrosoftSQLServer  enterprise  MicrosoftSQLServer:SQL2019-Ubuntu1604:enterprise:15.0.200114  15.0.200114
```

Figure 2.1: Ubuntu 16.04 images preconfigured with SQL Server

> **Note**
>
> There are multiple versions of SQL Server for a given major release such as SQL Server 2019 (15.0.x). Each corresponds to the build number of SQL Server. If your company has a requirement for a specific build of SQL Server and it matches one of these images, you can use it. If not, then you will need to utilize the second or the third method mentioned previously, either by deploying a VM without SQL Server pre-installed or by creating an image from scratch.

When using the Azure portal, the Azure Marketplace defaults to using the latest image version. This means that for older versions, such as the ones shown in *Figure 2.1*, you would need to use another method, such as PowerShell or CLI, to deploy that image.

The easiest way to find an image is to search for it in the Azure Marketplace in the Azure portal. For example, you can search for SQL Windows 2019 and filter further as desired, as shown in *Figure 2.2*. Note the limited choices as compared to *Figure 2.1*. In some cases, the image available will be a specific version and/or edition of SQL Server with an OS, while on other occasions, it will be a major version of SQL Server with an OS:

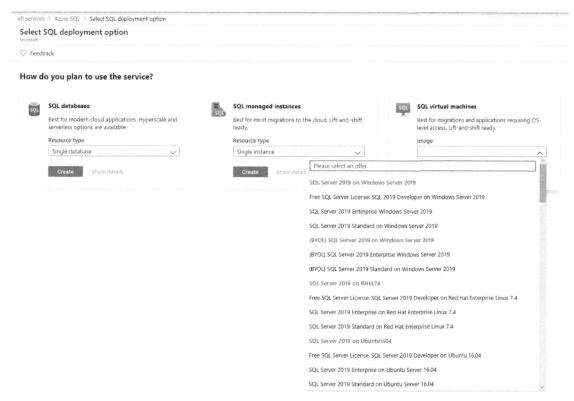

Figure 2.2: Windows Server 2019-based SQL Server images in the Azure Marketplace

Depending on the image selected, you may also have to specify the edition of SQL Server as shown in *Figure* 2.3:

Figure 2.3: Choosing an edition of SQL Server

The Azure portal allows you to configure a VM's settings manually when you click **Create**. You will be walked through a wizard via a series of panes where you configure storage, networking, and, for Windows Server-based configurations, aspects of SQL Server itself. Most IT organizations will choose this if using the Azure portal or automate.

You also have the option to **Start with a pre-set configuration**, as shown in *Figure* 2.3. An example of what that would look like can be seen in *Figure* 2.4. You still need to go through the rest of the configuration process, but what you would need to alter is much less. Full documentation can be found in the topics <u>Provision a Linux SQL Server virtual machine in the Azure Portal</u>[10] and <u>How to provision a Windows SQL Server virtual machine in the Azure portal</u>[11].

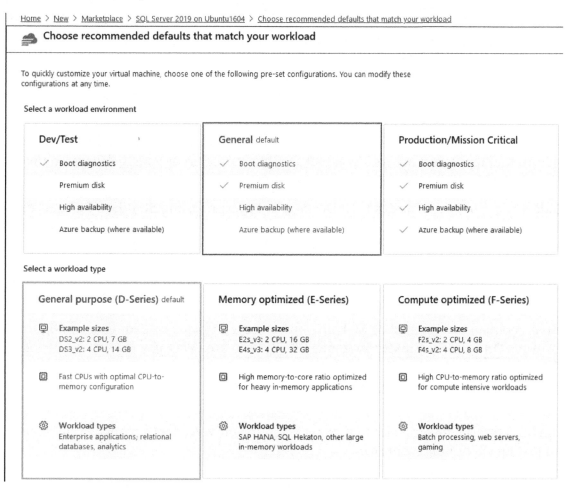

Figure 2.4: Choosing defaults

One difference to note is that, unlike a Windows Server-based SQL Server Azure VM image, there are fewer SQL Server options available during configuration. With Linux, you can only choose the core aspects of the VM. Anything related to SQL Server, such as changing default data paths, is configured inside the VM once deployed. More information on the differences compared with Windows Server can be seen in the upcoming section, SQL *Server on Azure Virtual Machines resource provider*.

For Linux, the image with SQL Server pre-installed contains the Database Engine, SSIS, and the command-line tools (**sqlcmd** and **bcp**). The end user license agreement (EULA) is already accepted. SQL Server Agent is not enabled. SQL Server Agent along with any other SQL Server features must be configured[12]. You will also need to reset the **sa** password, which is set during configuration and not published, and, if desired, add the tools to your default path.

This means, for Linux-based images, that you need to do some configuration after the VM is built, but not as much as if you had to install SQL Server yourself using an image with just the Linux distribution that is described in the next section.

Windows Server-based images with Developer, Enterprise, and Standard Editions come with many SQL Server options pre-installed: The Database Engine (including replication and R services, full-text search, and Data Quality Services), Analysis Services, and Master Data Services. If you want to change the configuration, the full installation media can be found on the local system drive. An example would be if you are not using Analysis Services; you can simply uninstall it.

Even if you plan on uninstalling SQL Server for reasons such as needing to configure an **Always On Failover Cluster Instance** (**FCI**), if you are using pay-as-you-go licensing (refer to the upcoming section on licensing for more information), using an image with SQL Server already installed gets you not only set up but also gets you the license(s) required.

If the Marketplace images with SQL Server do not suit your needs, consider an image with just an OS or build your own.

Using an image without SQL Server installed

Instead of using a template with the OS and SQL Server already installed, you can choose one that has the desired OS only. Choosing an image with just the distribution of Linux is similar to what was described in the previous section.

There are a few reasons why you may choose to install SQL Server yourself. Three example scenarios are listed here:

- An image has something desired, such as the high availability add-on in RHEL needed for both FCIs and AGs already installed.
- A pre-built template with your desired version of SQL Server does not exist for the target OS, which has an image. For example, as of the time of writing this chapter, no RHEL 8 template is pre-built with SQL Server 2019. Another example would be if your corporate standard was a specific version or a build of Linux such as RHEL 7.5, which is supported for SQL Server but there is no pre-built image.
- Your company has other specific requirements or challenges for a build that are not met by any existing image, but a base OS provides a starting point. An example would be SQL Server 2019 running on Windows Server 2016.

Figure 2.5 shows an example of selected RHEL images available in East US 2 that are only an OS or a special variant:

```
allan@Azure:~$ az vm image list --location eastus2  --all --output table --publisher RedHat --sku 8.1
Offer    Publisher    Sku    Urn                               Version
-------  -----------  -----  --------------------------------  --------------
RHEL     RedHat       8.1    RedHat:RHEL:8.1:8.1.2020020415  8.1.2020020415
allan@Azure:~$ az vm image list --location eastus2  --all --output table --publisher RedHat --offer RHEL-HA
Offer    Publisher    Sku    Urn                                  Version
-------  -----------  -----  -----------------------------------  --------------
RHEL-HA  RedHat       7.4    RedHat:RHEL-HA:7.4:7.4.2019062021  7.4.2019062021
RHEL-HA  RedHat       7.5    RedHat:RHEL-HA:7.5:7.5.2019062021  7.5.2019062021
RHEL-HA  RedHat       7.6    RedHat:RHEL-HA:7.6:7.6.2019062019  7.6.2019062019
```

Figure 2.5: Selected RHEL images

Once the VM with the chosen OS is built from the image, install SQL Server using the instructions found in the topic Installation guidance for SQL Server on Linux[13] or SQL Server installation guide[14] (for Windows Server-based VMs) and configure the instance of SQL Server to how you need it.

If a base OS image will not work, you still have one choice: create your own.

Creating your own image

To configure a custom hard drive image for use in Azure, follow the documentation linked below for each of the OSes supported by SQL Server:

- Red Hat Enterprise Linux (RHEL)[15]
- SUSE Linux Enterprise Server (SLES)[16]
- Ubuntu[17]
- Windows Server[18]

Generic information that applies to the three Linux distributions for creating a custom hard drive can be found in the documentation topic, Information for Non-Endorsed Distributions[19].

Similar to the previous section, you must also install and configure SQL Server in the OS.

SQL Server on Azure Virtual Machines resource provider

The SQL Server on Azure Virtual Machines resource provider is one way in which Microsoft makes deploying in Azure easier for administrators. This feature is only available for Windows Server-based VMs and can even be enabled if you deployed SQL Server yourself and did not use a Marketplace image pre-installed with SQL Server. To see how to perform this task, consult the following documentation: Register a SQL Server VM in Azure with the SQL VM resource provider[20].

The resource provider allows you to configure certain aspects of SQL Server, but, more importantly, also tasks such as backups that Microsoft can do for you instead of you having to configure it all inside the VM. An example is shown in *Figure 2.6*:

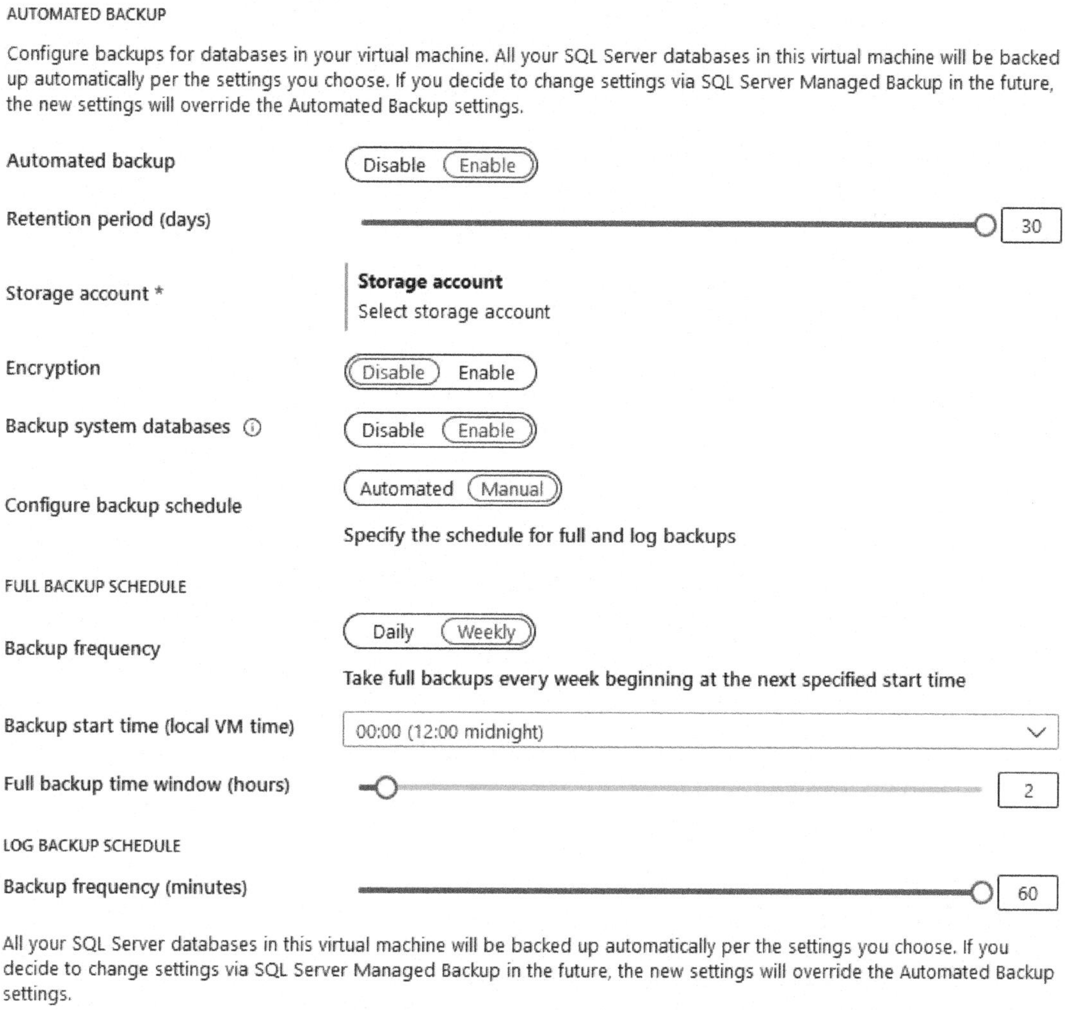

AUTOMATED BACKUP

Configure backups for databases in your virtual machine. All your SQL Server databases in this virtual machine will be backed up automatically per the settings you choose. If you decide to change settings via SQL Server Managed Backup in the future, the new settings will override the Automated Backup settings.

Automated backup	Disable Enable
Retention period (days)	30
Storage account *	**Storage account** Select storage account
Encryption	Disable Enable
Backup system databases ⓘ	Disable Enable
Configure backup schedule	Automated Manual Specify the schedule for full and log backups

FULL BACKUP SCHEDULE

Backup frequency	Daily Weekly Take full backups every week beginning at the next specified start time
Backup start time (local VM time)	00:00 (12:00 midnight)
Full backup time window (hours)	2

LOG BACKUP SCHEDULE

Backup frequency (minutes)	60

All your SQL Server databases in this virtual machine will be backed up automatically per the settings you choose. If you decide to change settings via SQL Server Managed Backup in the future, the new settings will override the Automated Backup settings.

Figure 2.6: Backups in the SQL Server resource provider

Other things that you can configure include security and patching. For patching, Microsoft will only apply updates marked as important, such as security updates. You will still have to apply things such as SQL Server Cumulative Updates, but you can specify the day, time, and maintenance window duration for applying important updates.

Common Azure VM deployment considerations for SQL Server

Last, but not least, there are some common considerations and best practices when deploying IaaS VMs in Azure for SQL Server, whether they are Linux- or Windows Server-based.

It is not recommended to assign a public IP address unless necessary as this exposes the server directly to the internet. VMs should be created on virtual networks that are private and accessible by authorized personnel. If connecting via on-premises to Azure, the assumption is that Express Route or a private VPN will be used so that the VM will be seen as if it was on-premises.

- If needed, open ports for accessing the VM, such as RDP (3389) or SSH (22).

- For most implementations of SQL Server, use a single virtual network interface card (vNIC), which is the default configuration. One vNIC is not a single point of failure as the underlying Azure network infrastructure is highly redundant and there are means such as Availability Sets and Availability Zones to ensure that VMs themselves will not be single points of failure.

- The VM relies on core elements of infrastructure such as DNS, to work properly. Even in a hybrid solution that would span on-premises and Azure, ensure that those key elements exist both on-premises and in the cloud.

Licensing SQL Server in Azure

Last, but not least, a major consideration for how you deploy in Azure relates to cost. All VMs deployed in Azure must be properly licensed even if what is inside is technically free. There are two aspects of licensing that must be accounted for: the OS and SQL Server.

Your choice of supported Linux distribution for SQL Server will dictate whether you need a paid license. For example, RHEL does require a license. All of the options are documented at the Red Hat on Microsoft Azure[21] page on Red Hat's website. Consult the *Red Hat Linux Enterprise Server* section at that link. Windows Server always needs to be licensed.

When it comes to SQL Server editions, Standard, Enterprise, and Web editions always require a license. SQL Server Developer and Express are technically free, but do have a license, and there may be restrictions regarding use. For example, Developer edition cannot be used for production workloads.

There are two models for licensing SQL Server in Azure: bring your own license (BYOL), or pay for the license as part of the cost of the VM, known as pay-as-you-go. For pre-built images with SQL Server, both options are often feasible. As of the time of writing this chapter, BYOL images are only an option for Windows Server-based SQL Server images.

If you are migrating to Azure or have unused licenses and also have Software Assurance, the Azure Hybrid Benefit for SQL Server allows you to use existing on-premises licensing and apply it to an IaaS VM. This can potentially reduce the cost of an IaaS VM.

There is a new Azure-specific licensing benefit introduced with SQL Server 2019 if you have Software Assurance: the ability to run a VM that has a standby server for disaster recovery. For example, if you have an on-premises Always On availability group and want to add an asynchronous replica in Azure, it is now free. This could represent a significant cost saving and can even be seen in the Azure portal, as shown in *Figure 2.7*:

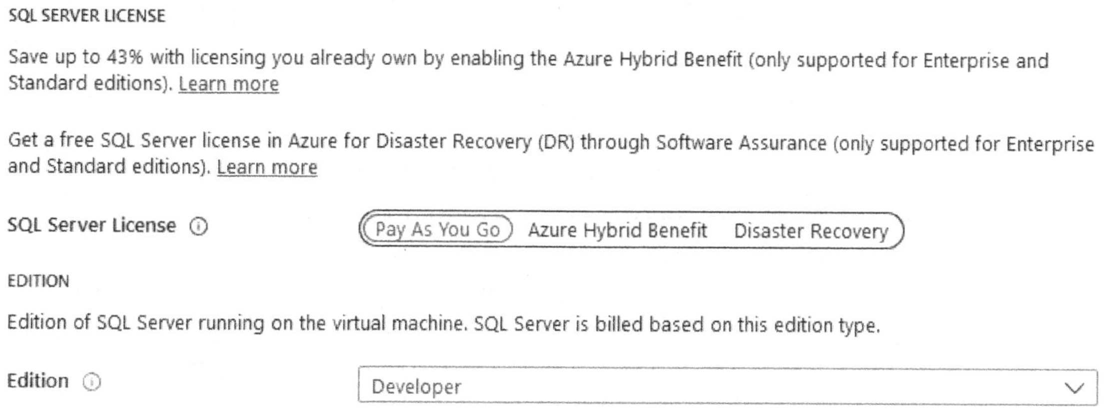

SQL SERVER LICENSE

Save up to 43% with licensing you already own by enabling the Azure Hybrid Benefit (only supported for Enterprise and Standard editions). Learn more

Get a free SQL Server license in Azure for Disaster Recovery (DR) through Software Assurance (only supported for Enterprise and Standard editions). Learn more

SQL Server License ⓘ   (Pay As You Go) Azure Hybrid Benefit   Disaster Recovery

EDITION

Edition of SQL Server running on the virtual machine. SQL Server is billed based on this edition type.

Edition ⓘ   Developer   ⌄

Figure 2.7: Licensing info from the Azure portal for a Windows Server-based VM

Full licensing information for SQL Server can be found on the Microsoft websites for on-premises pricing[22] and SQL IaaS pricing[23], as it will factor into the VM hardware choice and its cost since each SQL Server VM is usually licensed per virtual processor. More on choosing a VM size is covered in the next section.

Azure VM hardware options

VMs have a virtualized processor, memory, and storage. Processor and memory factor into the VM type and its choice of size. Storage is influenced by the VM type and size, but has its own parameters. This section will introduce the basics of VM types, sizes, and storage. Performance will be touched upon as this is a crucial element of configuring a VM and will be discussed further in *Chapter 5, Performance*.

VM types and sizes

This section will contain information about the different types and sizes of IaaS VMs available in Azure.

VM types and series

VMs in Azure come in different types and sizes. Within each major VM type, the size maps to what is known as a series, such as D, E, and G. Each series has different sizes with different specifications. The following table lists the different VM types and their purpose:

VM Type	Comments
General purpose	A good "all-rounder" VM but is not designed to stand out in any one manner. Everything is fairly balanced.
Compute optimized	These VMs have a high CPU-to-memory ratio (for example, an F series with 32 cores and 64 GiB of memory) meaning a higher density of compute power with a good amount of memory, but not a ton.
Memory optimized	These VMs have a high memory-to-CPU ratio. A VM with a lower vCPU count will most likely have more memory than its compute optimized counterpart. For example, an E-series with 20 vCPUs and 160 GiB of memory.
Storage optimized	These VMs are designed to emphasize better disk performance and will be discussed more in the next section, "Storage".
GPU	These VMs are meant for VMs that have high video or graphics use which would not be a SQL Server trait, so that is not a recommended type.
High performance compute	These are not recommended for SQL Server since Remote Direct Memory Access (RDMA) is not certified for in-guest storage in Azure.

Figure 2.8: Different VM types with their recommended usage

VM resources can be reserved and guaranteed by paying for a reserved instance. There is also the option to use dedicated hosts for the VMs, which would isolate performance further. Spot VMs allow you to use capacity in Azure, but if Azure needs those resources, the VM can be evicted. For that reason, a Spot VM is not recommended for permanent production SQL Server workloads. More information on Spot VMs can be found in the documentation in the Use Spot VMs in Azure[24] section.

VM size

Within each VM type category, there will be multiple VM sizes available, each with different capacities and limitations. Some VMs may not be available due to regional restrictions, subscription policies and constraints, quotas, and suchlike. The region(s) and VM sizes that you will be able to use may also depend on your company's standards as well as any policies they may put in place that could restrict how, what, and where things are done.

> **Note**
>
> Consult Microsoft Docs[25] to learn more about the details of the different VM types, the sizes currently available, the types of processors used for each, and their limitations.

How to choose a VM for SQL Server

Choosing a VM type and size depends on the database workload. This means that you need to know about the application and/or database profiles. Do you currently use a lot of CPU but not as much memory? Do you use more memory but not as much CPU? Do you pound your disks a lot on-premises and need a certain amount of guaranteed IOPS? A mixture of the above? These are the types of questions you will need to answer in order to pick the right type and size. The best way to know is to profile, baseline, and benchmark your application's workload to understand how it is using SQL Server and the underlying server.

Furthermore, a specific VM size will limit the amount of resources and the limitations are a hard cap. Once you hit it, that limitation cannot be upped with that VM size. To increase a limitation, the VM will need to be resized to a larger VM with minimal downtime.

Consider this example: as of the time of writing this chapter, there is a memory-optimized type VM size of Standard_E20s_v3/Standard_E20as_v4 that has 20 vCPUs, 160 GiB of memory, and up to eight virtual network cards with up to 8,000 Mbps speed. Storage aspects will be discussed in the next section.

A VM's number of vCPUs is static. Unlike using a hypervisor on-premises, what you select is what you get. If you need 17 vCPUs, the current minimum size for a memory-optimized VM that could be used would be the Standard_E20s_v3/Standard_E20as_v4. A Standard_E16s_v3/Standard_E16as_v4 has 16 vCPUs; it cannot be configured to have more. Read the descriptions in the link in the earlier side note for descriptions on the underlying CPUs, their speeds, manufacturers, and so on in order to make a correct determination of which type and size of VM to use. For example, while the Ev3 VMs are based on Intel processors, Eav4 VMs feature the AMD EPYC(TM) 7452 processor.

The same is true for memory. If you require 256 GiB of memory for a memory-optimized VM but only need 17 vCPUs, you must step up to the Standard_E32s_v3/Standard_E32as_v4, which has 32 vCPUs and the amount of memory required. You cannot add memory to an E20s/E20as. And if you require 2,000 GiB of memory but you only need 64 vCPUs, the Standard_Ea96s_v4 VM with 96 vCPUs would provide you with the right amount of memory. In both scenarios, to get more memory, a bigger VM must be selected.

The VM of choice must also account for the network throughput required. If you are planning on implementing a feature such as an availability groups for a busy database, you should know how much throughput you will need to ensure that the network will not become a bottleneck.

VMs can be resized to be bigger or smaller; however, downtime will be incurred. Plan accordingly.

Storage

This section will cover storage concepts for IaaS.

Disk types

There are five types of disks that a VM can use:

- Standard HDD
- Standard SSD
- Premium SSD
- Ultra Disk
- Temporary storage

Detailed information about the different types of disks and their limitations can be found in the documentation topic <u>What disk types are available in Azure?</u>[26], but their names are self-explanatory.

For SQL Server production workloads, Standard HDD and Standard SSD are generally not recommended. While they are less expensive, they often do not provide enough performance for demanding applications. Most SQL Server workloads will perform well on properly configured VMs using Premium SSD or Ultra Disk. Premium SSD also has a feature called Azure blob cache, which can improve performance. How it works is detailed in the blog post <u>Azure Premium Storage, now generally available</u>[27]. Ultra Disk is the fastest and most expensive storage, but may not be available in all regions or for all VM sizes. In addition, it may have to be enabled for the subscription. A sample message is shown in *Figure 2.9*, where there is some sort of restriction for Ultra Disk:

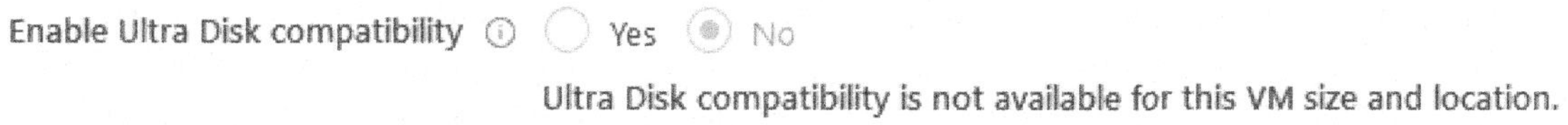

Enable Ultra Disk compatibility ⓘ   ◯ Yes   ⦿ No

Ultra Disk compatibility is not available for this VM size and location.

Figure 2.9: Ultra Disk not available for use

Sometimes, in order to attain the performance needed, multiple disks need to be configured and then, inside the VM, grouped together with Storage Spaces (Windows Server) or using the Linux tools to create a single logical volume from two physical disks presented.

Disk capacity

Disk types are one piece of the storage puzzle. The other is size, or capacity. Each disk type has different sizes of disks to choose from. For example, as of the time of writing this chapter, Premium SSDs have sizes with names such as P30 and P40. Each one has different maximum specifications. For example, a P30 disk today has a maximum size of 1,024 GiB and a P40 2,048 GiB.

If your database size and projected growth exceeds the smaller size, but is less than the next one, you will need to purchase the larger size. This means that if your database is over a terabyte in size (a P30), but less than two (a P40), you will need to consider a P40 if you wish to configure a single disk. There are other ways to achieve capacity greater than a terabyte that will be described in the following sections.

Storage performance

Storage performance is different, but a completely related concept. Part of choosing the right disk configuration is understanding your performance requirements. Also similar to VMs and deciding what type, series, and size to choose for the right processor and memory, you must choose your disk configuration based on the performance required because the limits are rigid. The preceding link documents accurate guidelines for the performance you can expect from each type of disk in certain categories.

Not all parameters are documented for the different disk types. For example, a disk rated at a specific speed will deliver up to that for whatever kind of I/O was tested, but it may not work well with your workload. Always test to ensure that you are getting the performance required before going into production with your workload.

Consider this example: as of the time of writing this chapter, a single P30 can sustain up to 5,000 IOPS at 200 MiB/sec, and the larger P40 disk, 7,500 IOPS at 250 MiB/sec. If you have a database that requires half a terabyte of space but needs 15,000 IOPS or a sustained 450 MiB/sec, you are possibly looking at the equivalent of a P60 if using a single disk or configuring multiple smaller disks (either to look like one disk or spreading the database across those disks).

There is a second and equally important aspect that gates storage performance: the VM type and size. Each VM is rated for storage throughput. A disk with higher throughput ratings can be attached to a VM that has lower throughput, but the storage will only run as fast as allowed by the VM type and size.

Assume that you are using a Standard_E20s_v3 VM. For storage, it can have a maximum of 32 data disks, 320 GiB of temporary SSD storage (which has a ceiling of 40,000 IOPS/320 MBps of throughput with 400 GiB of cache), and a maximum of 32,000 IOPS/480 MBps of throughput for the VM outside of temporary storage.

If your current production workload needs nearly a gigabyte per second of throughput, the E20s will not do it. Assuming you want to stick with an E-series VM, you would be looking at a size of E48s or E64s. The limit of 480 MBps for the E20s is hard; the VM will never achieve more. Looking at the choice of disk, a single P60 would come in at half the amount of IOPS that an E20s VM could achieve. The VM would not let it achieve more than that figure of 480 MBps. This means that the single P60, rated at 500 MBps, could never hit that. You could use two P80 disks to achieve 40,000 IOPS, but would still be capped at a maximum throughput of 480 MBps, even though each P80 disk is rated at 900 MiB/sec. Choosing a VM hardware is about making trade-offs and compromises. Similar considerations would apply to the EAs VM series, even if the throughput information you can find in the documentation will be different.

Another restriction is that certain VM sizes cannot use premium disk types, which also may factor into what VM you use. If you're using the Azure Marketplace, you will see a message similar to the one in *Figure 2.10*:

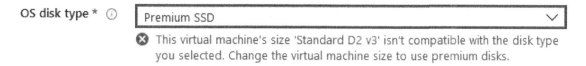

Figure 2.10: Error message when trying to use Premium SSD

If you are using a Windows Server-based Marketplace image pre-installed with SQL Server, you can select the type of workload and get some assistance optimizing the storage configuration, as shown in *Figure 2.11*:

Configure storage

Storage optimization ⓘ   (General (Transactional processing)   Data warehousing)

Data storage

These disks will be attached to your virtual machine as data disks and will be stored in storage as page blobs.

Data drive location * ⓘ   **Disk type** * ⓘ

| F:\data | ✓ | | Premium SSD | ⌄ |

Disk type	Size (GiB)	Max IOPS	Max throughput	Number of disks
1024 GiB, Premium SS... ⌄	1024	5000	200	1

ⓘ 1024 GiB, 5000 IOPS, 200 MB/s

Log storage

Transaction logs are a critical component of the database as they record all transactions and database modifications made by each transaction.

Shared drive space * ⓘ   **Log drive location** * ⓘ   **Disk type** * ⓘ

| Use a separate drive for l... ⌄ | | G:\log | ✓ | | Premium SSD | ⌄ |

Disk type	Size (GiB)	Max IOPS	Max throughput	Number of disks
1024 GiB, Premium SS... ⌄	1024	5000	200	1

ⓘ 1024 GiB, 5000 IOPS, 200 MB/s

TempDb storage

The tempDb system database is a global resource that is available to all users connected to the instance of SQL Server. It is used to store temporary user objects and internal objects created by the database engine.

Shared drive space * ⓘ   **TempDb drive location** * ⓘ

| Use local SSD drive | ⌄ | | D:\tempDb |

⚠ The desired performance might not be reached due to the maximum virutal machine disk performance cap. The selected VM size (Standard_D2s_v3) only supports up to 3200 disk max iops (currently 10000 iops), 48 disk max throughput in MBps (currently 400 in MBps).

❌ The selected VM size (Standard_D2s_v3) only supports up to 4 data disks (currently 2 data disks).

| Ok | | Discard |

Figure 2.11: Storage configuration pane in the Azure portal

Note that what has been discussed previously is also seen at the bottom in the warning where you may not get the throughput needed and the limitation of the number of disks. Another good aspect to being able to configure this at the time of provisioning the VM is that you can enforce standards for things such as drive letters and data and transaction log file folders.

Ephemeral storage and SQL Server data and transaction log files

Each VM has temporary, also known as ephemeral, storage. Anything configured on this storage is lost if the VM is shut down or rebooted. Therefore, it is not recommended for SQL Server data or transaction log files for application or user databases.

The only potential use for ephemeral storage is TempDB. TempDB is recreated every time SQL Server is restarted, so by the nature of its design, what is in it is not permanent.

There is one caveat if you choose to use this storage for TempDB: the size of the ephemeral disk is fixed and can never be expanded. The only way to make it bigger is to choose a different VM size. That also means that performance cannot be greater than what the VM allows for that disk without resizing. If you know that your TempDB usage meets size and performance requirements, you can consider ephemeral storage since it generally performs better for some things, including 8 KB writes.

Storage-optimized VMs use local non-volatile memory express (NVMe) storage that is ephemeral. If the VM is rebooted, anything configured on the temporary disk is lost. This means that storage-optimized VMs as they are configured as of the time of writing this chapter are not recommended for SQL Server use.

Summary

Choosing a VM type, size, and its associated virtual hardware is no different to planning and deploying a physical server or VM on-premises. You have to account for CPU, memory, networking, and storage. While this chapter covered the basics of deploying an IaaS-based VM for SQL Server in Azure, there are many more considerations, such as availability and security, that must also be considered as part of an overall solution. Subsequent chapters will discuss those topics and more.

Chapter links

1. https://bit.ly/2ZFMfx7
2. https://bit.ly/3d4oiDw
3. https://bit.ly/3gkCUkc
4. https://bit.ly/2X1burV
5. https://bit.ly/2LXSYdK
6. https://bit.ly/2TB4sHY
7. https://bit.ly/2A4dsP6
8. https://bit.ly/2zi0ai5
9. https://bit.ly/2zi0hdv
10. https://bit.ly/36rGrlE
11. https://bit.ly/3d1Xyn4
12. https://bit.ly/2ZBerkK
13. https://bit.ly/2Xuj0dI
14. https://bit.ly/2TDBFTh
15. https://bit.ly/2A4evyw
16. https://bit.ly/2XpdV6s
17. https://bit.ly/3ei9MZ2
18. https://bit.ly/2LXkA2p
19. https://bit.ly/2WXTFde
20. https://bit.ly/3bXoIKx
21. https://red.ht/2A8ltRQ
22. https://bit.ly/2yuFhj5
23. https://bit.ly/3eeClq7
24. https://bit.ly/2ZBghlE
25. https://bit.ly/2LTjvJ7
26. https://bit.ly/3ekIpNZ
27. https://bit.ly/2LVFCOZ

3

Hero capabilities of SQL Server on Azure Virtual Machines

By Joey D'Antoni

While **platform as a service (PaaS)** databases such as Azure SQL Database have many upsides, the most common SQL Server deployment method on Azure is in an IaaS **virtual machine (VM)**. The typical reasons for this are that partner software may not support PaaS options, or that you may need to work on older versions of the SQL Server engine. Another common scenario is when an organization chooses to run another SQL Server component, such as **SQL Server Integration Services (SSIS)**, alongside the database engine. Some of the other included features that may be used include PolyBase, which allows data virtualization to other data sources such as MongoDB, Oracle, or Teradata. Also, Machine Learning services allows users to execute R, Python, or Java scripts side by side with SQL Server. While this architecture may not offer the best performance, it does maximize the value of licensing. In this chapter, you will learn about:

- Configuring your VMs to be highly available.

- Monitoring performance over time.

- Optimizing disk layout for SQL Server on Azure VMs

- Benefits of SQL VM resource provider

- Managing your SQL Server estate in Azure.

- Using Azure Active Directory managed identities.

- SQL Server 2019 security features.

Understanding platform availability in Azure

It is important to understand that the Azure infrastructure is built and designed to be highly available, but just like in every computer system, there are failures that can happen. Azure offers you a couple of different ways to build resiliency within your infrastructure. Azure is made up of regions and spread across geographies, as shown in *Figure* 3.1:

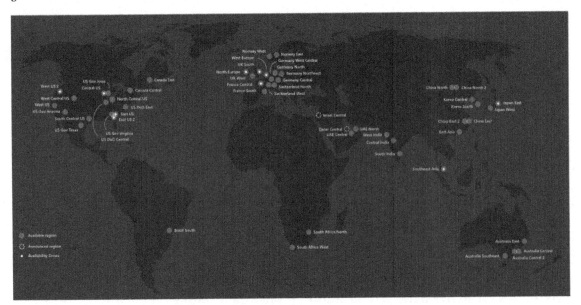

Figure 3.1: Azure regions throughout the world

A geography is a designation that ensures that data residency respects geopolitical boundaries and meets any data sovereignty. A geography is typically defined by the borders of a country (typically Microsoft won't add a region without adding two regions in a given country); however, there are a few exceptions, such as Brazil South, which is a single region within a single country.

Also, Azure will always define regional pairs within its infrastructure. This is an important concept in terms of the overall availability of the platform. Azure paired regions have a few important concepts to take note of:

- If you are using geo-redundant storage, it will be replicated to the paired region (you don't get to choose the target).

- Microsoft performs updates of the infrastructure in a serial fashion across paired regions so that in the event of an update failure, the failure will not cascade.

- Paired regions are at least 500 kilometers/300 miles apart where possible so that they are protected from natural disasters.

- Finally, in the event of a major Azure outage, Microsoft will prioritize the recovery of one region out of every pair.

If you are using IaaS, you do not have to deploy your resource to paired regions for disaster recovery, but it may be in your best interest to do so as it might provide you with the highest levels of availability.

While each region is made up of multiple physical datacenters, the lowest level of granularity you have when you are deploying Azure services is the region. This means that in the event of a physical outage in a datacenter, your application could incur downtime if you do not have a disaster recovery solution in a second region.

Availability Zones

In 2017, Azure introduced Availability Zones, which allow you to split workloads between physical datacenters. When you deploy a VM resource to an Availability Zone, you have the choice of deploying to zone 1, 2, or 3, which can spread your workloads across multiple datacenters in a given region, as shown in *Figure* 3.2:

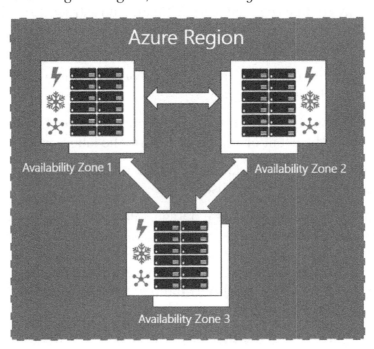

Figure 3.2: Availability Zones

Availability Zones have separate power, network connections, and cooling in order to prevent a single physical failure from taking down your workloads. While *Figure* 3.2 and the portal will always present you with three logical Availability Zones, in actuality each region that has Availability Zones is divided up into four physical zones—the zones you choose in the portal while ensuring that your workload is placed in different datacenters within the region are not tied to a **specific** datacenter every time. For example, my deployment to Zone 1 in the East US 2 region may not be in the same datacenter as your deployment to Zone 1 in the same region.

While network latency between datacenters in a region is low, it is important to test that latency before deploying your workloads, as some regions will have higher latency between zones than others. This could affect SQL Server when deploying a technology such as Always On availability groups; you might choose to use synchronous replication if the latency is less than one millisecond for a critical transaction processing application. On the other hand, if the latency is higher, you might risk some data loss with asynchronous replication. It is very important to test the latency between your VMs before deploying your architecture.

Availability sets

In addition to Availability Zones, Azure also has availability sets, which provide availability within a datacenter within a region. A simple way of thinking about this is that if you have three VMs in an availability set, they are all deployed to different racks within the datacenter. In reality, it is more complex than that, with Azure being broken down into update domains, which allow the infrastructure to be updated, and fault domains, which provide isolation to single points of failure in the core infrastructure. This, combined with managed disks, offers 99.95% uptime for multi-VM deployments.

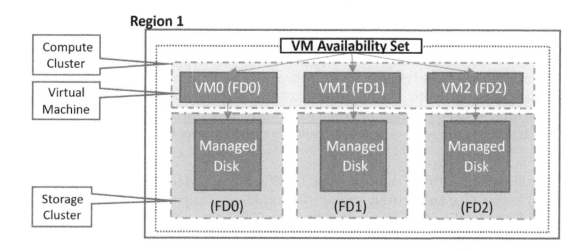

Figure 3.3: Availability set configuration

If you are deploying Always On availability groups or Failover Cluster Instances, you will need to deploy them into an availability set or into an Availability Zone. This is required in order to configure the internal load balancer, which will act as the IP address for the availability group's listener (the virtual IP address that users and applications use to connect directly to the availability group).

While all these solutions provide high availability, none of them protect against regional disasters. One of the benefits of using the public cloud is the ease of putting workloads in multiple regions to protect against natural disasters or regional failure.

Disaster recovery options for SQL Server in Azure

The first step in any good disaster recovery plan is having reliable and redundant backups. SQL Server and Azure work together to make this simple—SQL Server 2012 (specifically Service Pack 1, Cumulative Update 2) onward has supported backing up directly to Azure Blob storage through the **BACKUP TO URL** syntax. This feature was enhanced in SQL Server 2016 and supports backups larger than 1 TB, and striping the backup to improve performance. Commonly used tools such as the built-in maintenance plans and Ola Hallengen's maintenance scripts support backing up to Azure Blob storage.

> **Note**
>
> While backup to URL is supported in tools, you should note that neither maintenance plans nor Ola Hallengren's scripts support the pruning of older backups. A common workaround for this is to add an SQL Server Agent job step that removes the older backup files upon completion of your backup tasks.

Additionally, you have to use Azure Backup for SQL Server, which builds on top of the Azure Backup service and actively manages your older backups. Azure boosts your availability by providing **Geo-Redundant Storage** (**GRS**) accounts, which provide two copies of each backup file stored in a highly available fashion (Azure Storage doesn't use traditional RAID patterns, instead requiring three copies of a file within a region for a write to be considered complete) in two different regions. This replication is asynchronous, which makes it well suited for backups, but not for the storage of SQL Server data files.

Beyond backups

In addition to backups, you should capture your configuration's ARM templates and any SQL Server configuration that exists outside of your databases (cluster configuration, quorum drives) in your source control system. This provides for budget-conscious customers who do not want to have a real-time secondary replica of their database.

Another low-cost (and simplified configuration) option is to use Azure Site Recovery to protect your workloads. Azure Site Recovery replicates your VMs using block-level storage replication from one Azure region to another. There is no need for the target region to be online, which helps reduce the compute costs. Your recovery data is stored in a storage vault in the secondary regions. Azure Site Recovery is priced at $25 USD per instance per month (an instance in this scenario is a VM).

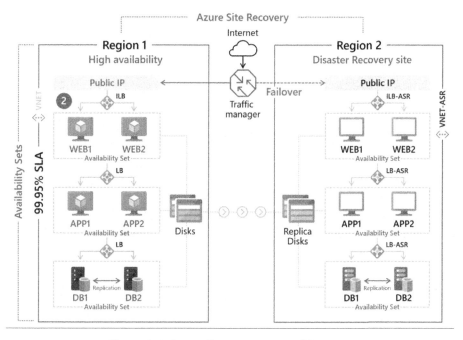

Figure 3.4: Azure Site Recovery Architecture

Azure Site Recovery is not the disaster recovery solution for all workloads—you could potentially lose somewhere between 5 and 60 minutes of data, depending on the failure mechanism. However, it is very cost effective and very simple in terms of setup and management. For high availability and disaster recovery solutions with a lower **Recovery Time Objective (RTO)** and **Recovery Point Objective (RPO)**, most customers are going to use either Always On availability groups or log shipping techniques.

Always On availability groups

Introduced with SQL Server 2012, Always On availability groups provide both high availability and disaster recovery by allowing synchronous and asynchronous replication between one or more database(s). Additionally, availability groups allow reads to take place on the secondary replicas, allowing you to scale your architectures to put replicas closer to end users or behind a load balancer for general reporting. In SQL Server 2019, you can have up to five synchronous replicas in your availability group (and up to eight total replicas). Synchronous and asynchronous in this context refer to the process of transaction hardening. When an availability group is running in synchronous mode, a transaction is not considered complete on the primary replica until it reaches (and is hardened into) the transaction log of the secondary replica(s). In most configurations, the transaction is considered committed when it reaches the first replica; however, there is an optional `required_copies_to_commit` setting that provides additional data protection by ensuring the database transaction reaches additional replicas.

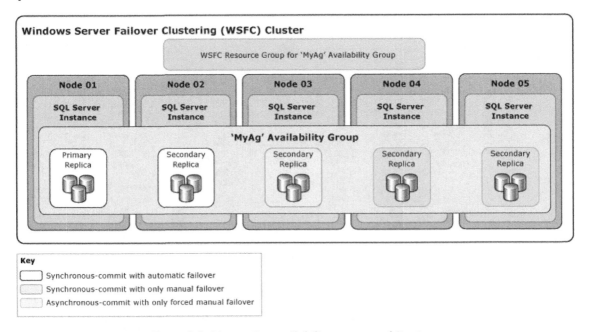

Figure 3.5: Always On availability group architecture

Synchronous replication is typically used when the servers are in the same physical location. In Azure, this would mean that if your VMs are in the same Azure region, synchronous replication would be your choice. The introduction of Availability Zones does add some complexity to this design, as the network latency between zones, while typically low, can differ in different Azure regions. There is no hard-and-fast rule for when to choose synchronous over asynchronous replication, but you should typically consider 10 milliseconds of latency to be the absolute upper limit for synchronous replication. Depending on the nature of your application and its sensitivity to latency (for example, a stock trading application would be extremely latency sensitive, whereas a business intelligence reporting system could tolerate some levels of latency), you may require much lower latency than that.

The other component of synchronous replication is that it is required for automatic failover. SQL Server will not allow you to lose data without accepting the potential data loss, and any asynchronous replication solution will allow you to lose transactions that have committed to the primary database but have yet to reach the transaction log on the secondary database.

While availability groups have typically been deployed in conjunction with Windows Server failover clusters, if you are running SQL Server on Linux it uses Pacemaker as its clustering component. You can learn more about configuring SQL Server with Pacemaker here[1]. Finally, you can build an availability group with no clustering software; however, this architecture is typically used as a read-scale option, as it does not provide the same levels of automated failover that a clustering solution would.

Differences with availability groups in Azure

In general, configuring SQL Server VMs in Azure is very similar to on-premises VMs or even physical servers. There are a couple of differences, which will both be covered in this chapter–storage and networking. Availability groups use a listener, which is an alias, and an IP address to direct connections into the availability group. The listener will route traffic to the primary replica of the group and, optionally, route read-only traffic to replicas. The listener relies on gratuitous **Address Resolution Protocols (ARPs)**, which can broadcast MAC addresses to IP addresses, which reassigns the IP in the event of failover. Since Azure virtual networks do not support gratuitous ARPs (they do not allow broadcasting for security purposes), you need another mechanism to assign a "floating" IP address.

> **Note**
>
> Azure provides a number of ways to deploy VMs. You have the option to use PowerShell, the Azure portal, the Azure command-line interface (CLI), or ARM templates. You can quickly get started with an availability group by using the quickstart templates[2].

In Azure, this floating IP addressed is managed by the **internal load balancer (ILB)**, which acts as a front end for your availability group listener (and can also act as a front end for your clustering software). A load balancer has a relationship with your virtual network, and then a back-end pool of targets, which in an availability group scenario are your SQL Server instances. There are some differences in ILBs, depending on whether you are using availability sets or Availability Zones as your protection mechanism. If you are using sets, you can use a **Basic** load balancer, while Availability Zones require a **Standard** load balancer. You can find the steps to create a load balancer at this Microsoft documentation[3]. For SQL Server on Linux, you will not need to run the PowerShell steps to register the listener into the cluster, but you will need to associate the Pacemaker cluster with the IP address of the listener.

The additional complication for availability groups is the disaster recovery architecture that spans Azure Virtual Networks. This configuration is referred to as a **multi-subnet availability group**, and it is a fairly common on-premises configuration. In Azure, you simply have to create an ILB for each virtual network where you have availability group replicas. There are also some recommended DNS changes for this type of configuration. You can read more about how to configure for multiple subnets here[4].

Availability groups for read-scale workloads

One of the key capabilities of Always On availability groups, in addition to offering a high availability/disaster recovery solution for SQL Server, is the provisioning of readable secondary copies of the database. This can be used in conjunction with read-only routing and the load balancing functionality in SQL Server to spread traffic over multiple readable secondary replicas. You can use this functionality to place copies of the database closer to your end users, or to offload the reporting of the primary replica and add secondary copies locally.

SQL Server on Azure VM resource provider

When you deploy an SQL Server VM from Azure Marketplace, as seen in *Figure* 3.6, part of the installation process is the IaaS Agent Extension.

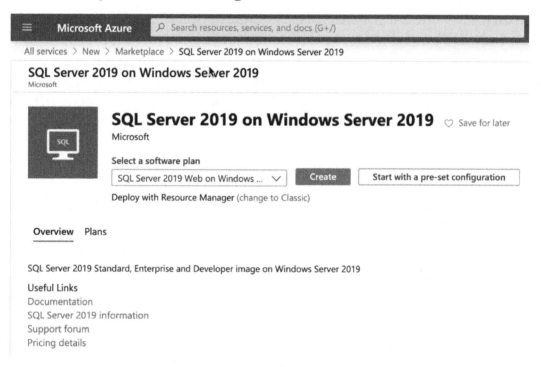

Figure 3.6: SQL Server VM creation from Azure Marketplace

Extensions are code that is executed on your VM after deployment, typically to perform post-deployment configurations, such as installing anti-virus or installing a Windows feature. The SQL Server IaaS Agent Extension provides four key features that can reduce your administrative overhead:

- **SQL Server Automated Backup**: This service automates the scheduling of your backups on the VM. The backups are stored in Azure Blob storage.

- **SQL Server Automated Patching**: This VM setting allows you to configure a patching window in which Windows updates to your VM can take place. Only SQL Server updates that are pushed down through the Windows Update process will be applied. At the time of writing, that is limited to SQL Server GDR updates, which means in order to keep your SQL Server VM fully patched, the DBA needs to install cumulative updates to SQL Server.

- **Azure Key Vault Integration**: This integration enables you to use Azure Key Vault as a secure storage location for SQL Server certificates, backup encryption keys, and any other secrets, such as service account passwords.

- **License Mobility**: You can change your license type and edition of SQL Server and switch from pay-as-you-go to pay per usage, **Azure Hybrid Benefit (AHUB)** to use your own license, or disaster recovery to activate the free disaster recovery replica license.

In addition to these features, the extension allows you to view information about your SQL Server's configuration and storage utilization, as shown in *Figure 3.7*:

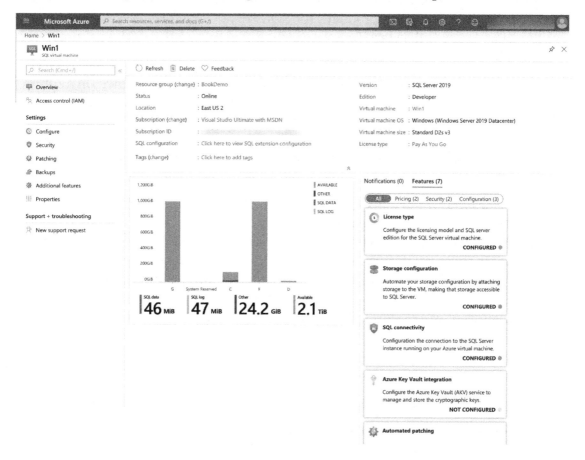

Figure 3.7: SQL VM configuration in the Azure portal

Performance optimized storage configuration

VMs registered with SQL VM resource provider can automate storage configuration according to performance best practices for SQL Server on Azure VMs through the Azure portal or Azure Quickstart Templates when creating an SQL VM. Best practices are detailed below:

- Separating data and log files to different volumes makes a difference on Azure because data and log files have different caching requirements. This feature can be automated when storage is configured through the Azure portal or Azure Quickstart Templates. Hosting data and log files on the same drive is supported only for general-purpose workloads; separate drives are the default configuration for OLTP and DW workloads.

- The performance of TempDB is critical for SQL Server workloads because SQL Server uses TempDB to store intermediate results as part of query execution. The local storage (D: drive) available to Azure VMs has very low response times and is included in the cost of the VM. Hosting TempDB on local storage has significant price/performance advantages if the size and the storage scale limits of the VM are sufficient for the workload. Measure the I/O bandwidth needed to meet the demands of workloads and test to find the required storage capacity for the TempDB. If the local storage capacity on the VM is not enough for the workload's TempDB requirements, consider hosting TempDB on Premium SSD or ultra disks to get very low response times. Performance-optimized storage configuration automates hosting TempDB on the local storage of the VM. SQL VM resource provider automates the re-configuration required after a restart, allaying concerns about failovers and VM restarts. Hosting TempDB on the local disk is the default configuration for OLTP and DW workloads and is supported for general-purpose workloads.

- Azure ultra disks deliver high throughput, high **I/O Operations Per Second (IOPS)**, and consistent low-latency disk storage for Azure VMs. When storage latencies bottleneck, use ultra disks to increase the throughput. Premium disks have great price/performance advantages with read-only caching and a low-end monthly storage cost. If workloads require storage response times at the microsecond level, use ultra disks as they provide consistent sub-millisecond read and write latencies at all IOPS levels (up to 160,000 IOPS). Leverage ultra disks to optimize storage performance for log files or TempDB files (if the local disk on the VM does not have enough capacity). For read-heavy TPC-E type workloads with limited data modifications, increase throughput by hosting data files on ultra disks. Performance-optimized storage configuration supports using ultra disks to host data, log, and TempDB files through the Azure portal. Azure Quickstart Templates can also deploy an SQL VM with a log file on an ultra disk.

- SQL Server images on Azure Marketplace come with a full and a default installation of SQL Server. SQL Server Database Engine-only images work for SQL Server 2016 SP1, SQL Server 2016 SP2, and SQL Server 2017 Enterprise and Standard editions. Those images can be used to create an SQL VM through the Azure portal, PowerShell, or ARM template deployments. Use the free manageability to simplify SQL Server administration and the performance-optimized storage configuration to boost SQL Server performance on Azure VMs by creating a new SQL VM through the Azure portal or by registering with SQL VM resource provider today.

SQL Server performance in Azure VMs

Many customers are concerned about how their critical workloads will perform after migrating to the public cloud. Given the multitude of VM types available within Azure, there is an extremely wide range of performance options. You can build a VM that's as small as 1 CPU and 0.75 GB of RAM all the way to 416 vCPUs and 12 TB of memory. Beyond that, each VM has a specific limit on storage and network bandwidth and the number of IOPS that the VM can perform. It is important when you are planning a migration to monitor your on-premises workloads so that you can make your Azure footprint the right size. This is particularly important for a **relational database management system (RDBMS)** such as SQL Server, which is I/O and memory intensive and does not offer easy horizontal scale options such as a web or application tier. Typically, if you have to increase the performance of SQL Server, you have two choices: purchase more hardware or optimize your queries.

One complication of the public cloud is the number of CPU cores aligned with the amount of RAM in a given VM. SQL Server is an application that is heavily dependent on RAM for its performance—throughout the database engine, memory is used to prevent calls to disks that are orders of magnitude more expensive. While having additional vCPU cores will not harm performance in an Azure environment, because SQL Server is licensed by the core, it can create a great deal of additional expense. Microsoft has identified this as an issue and offers a number of constrained core VMs for database workloads. You can identify a constrained core VM by its nomenclature—for example, in Standard_M64-16ms, M identifies the VM class, 64 identifies that this VM would typically have 64 vCPUs, and the -16 indicates that the VM is constrained to 16 cores. The full list of constrained core VMs is available here[5]. The compute costs for these VMs are the same as if they were unconstrained, but you are not responsible for licensing SQL Server for those non-allocated cores.

Azure Storage

While compute sizing is relatively straightforward, building Azure Storage for performance is slightly more complex. Azure VMs should use managed disks for availability and ease of configuration. Azure offers four types of managed disks to meet your performance and budgetary requirements:

- Standard storage

- Standard SSD

- Premium storage

- Ultra disks

Standard storage will not meet the performance requirements of SQL Server data and log files—it is useful for backing up SQL Server databases, along with any typical file storage that does not require low-latency access. Likewise, standard SSD provides similar but more consistent performance than standard storage, and should not be used for workloads as I/O intensive as SQL Server.

This leaves you with premium storage and ultra disks as your options for database storage. Premium storage typically provides single-digit latency and is particularly effective for read workloads, as it can take advantage of read-caching to provide even better performance.

> **Note**
>
> Azure VMs have a local SSD that is mounted on /dev/sdb on Linux, and the D:\ drive on Windows. This disk is ephemeral, and data on it may be lost during maintenance activities or when you redeploy a VM. It may be used for TempDB in conjunction with SQL Server, but you should note latency because if you are using read-caching for your disks, the read cache will exist on that temporary drive and may cause contention with your TempDB performance.

Premium storage requires more configuration to approach the performance of ultra disks. A common example is where a database has 2 TB of data but requires 30,000 IOPS. To meet the data volume requirement on premium storage, you could simply allocate one P40 disk, which has 2 TB but only offers 7,500 IOPs. To meet the performance requirement, you should consider allocating six P30 disks, which would have a total of 6 TB of storage but meet the IOPS requirement. In order to achieve this amount of IOPS with premium storage, you would stripe your data volumes across the six disks, giving you a volume of 6 TB and meeting the 30,000 IOPS requirement. You can read more about how to configure this on a Linux VM[6] and for a Windows VM[7] at the respective Microsoft documentation links. You should note that there is no requirement to mirror the disks in an Azure configuration because Azure provides redundancy at the infrastructure level in order to provide data protection. You should also note that each VM type has a specific amount of IOPS and storage bandwidth available to it, and throttling will kick in as the amount of IOPS approaches that threshold. This applies to all storage types, including ultra disks.

Ultra disks simplify this configuration but are more expensive than premium storage. Rather than simply paying for each disk based on volume and IOPS, the ultra disk option charges for each component individually. You choose the volume of the disk, the amount of IOPS, and the amount of bandwidth for the disk, and the cost is the total of each of those components. This simplifies the configuration, as you can create a single disk that meets your capacity and performance requirements. Ultra disks do not provide caching but can provide performance in the sub-millisecond range for some workloads and have the most consistent performance under heavy load of any of the storage options in Azure. The ultra disk option is built using NVMe storage and **remote direct memory access (RDMA)** to deliver this level of performance.

Disk layout for SQL Server on Azure

When designing a storage architecture for SQL Server, you should first think about the ways SQL Server performs I/O. For example, when you write an insert or update statement, the following activities happen:

1. The data page where the write takes place is updated in memory.

2. The insert or update is directly written to the transaction log.

3. The transaction is marked as complete.

4. Eventually, the data page in memory is flushed to disk, either via the SQL Server lazy writer process or via a checkpoint.

As you can see in this example, the most important factor in completing the transaction is how quickly the write to the transaction log takes place. Another scenario is a query that needs a great deal of memory to execute a join operation. SQL Server's behavior is that if the amount of memory it needs is not available, it will spill into TempDB, effectively treating TempDB like a page file. So, just like the transaction log, it is important for TempDB's data files to have extremely low latency to meet performance requirements.

To translate this to Azure Storage, to maximize performance and minimize costs, you might consider provisioning an ultra disk to host your transaction log and TempDB files, and premium storage with read-caching to store your data files. If your latency requirements are lower, you might consider creating two volumes—a premium storage volume with read-caching enabled, and another premium storage volume without caching for transaction logs and TempDB.

Backups

Backups are critical in terms of data availability and, as mentioned earlier, SQL Server supports directly backing up databases into Azure Blob storage. The basics and troubleshooting guide for this process are in this <u>document</u>[8], but we should highlight a couple of basics. When backing up to Azure (or in nearly all on-premises scenarios), you should use the **WITH COMPRESSION** option because it reduces the size of your backups and shortens both the backup and restore time. If the size of your backups exceeds 1 TB, you will need to stripe your backup across multiple files and adjust the **MAXTRANSFERSIZE** and **BLOCKSIZE** options, as shown in the following example:

```
BACKUP DATABASE TestDb

TO URL = 'https://mystorage.blob.core.windows.net/mycontainer/
TestDbBackupSetNumber2_0.bak',

URL = 'https://mystorage.blob.core.windows.net/mycontainer/
TestDbBackupSetNumber2_1.bak',

URL = 'https://mystorage.blob.core.windows.net/mycontainer/
TestDbBackupSetNumber2_2.bak'

WITH COMPRESSION, MAXTRANSFERSIZE = 4194304, BLOCKSIZE = 65536;
```

In addition to backup to URL, SQL Server supports using Azure Backup, which can automatically manage your database backups across all of your Azure VMs. Azure Backup installs an extension called **AzureBackupWindowsWorkload**, which manages the backup using a coordinator and an SQL plugin, which is responsible for the actual backup.

Gathering performance information

SQL Server is extremely well-instrumented software and offers you a number of ways to gather performance data. It uses extensive dynamic management views and system catalog views that let you retrospectively gather data on the performance of the database engine. The extended events engine in SQL Server allows you to trace code execution and isolate specific activities. The Query Store, which was introduced in SQL Server 2016 and has been continually improved since, allows you to capture runtime and execution plan information about individual queries to isolate any change in performance caused by volume changes or parameter changes.

Dynamic management views are covered in detail in *Chapter 5, Performance.*

Query Store

The Query Store feature has a number of benefits in that it acts as a flight data recorder for SQL Server. It does have to be enabled, and this action can be performed either through T-SQL or using **SQL Server Management Studio (SSMS)** to change the **Operation Mode** to **Read write**, as illustrated in *Figure 3.8:*

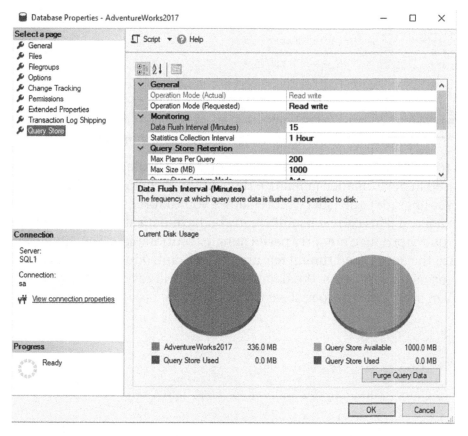

Figure 3.8: Query Store configuration

The Query Store is configured in each user database; busy databases or databases that have a lot of dynamic SQL (and therefore unique query strings) can require more storage. Query Store data exists in the primary filegroup for the individual data so that it is persisted across server restarts and availability group failovers.

There are also a number of reports built into SSMS that allow you to look at various perspectives of query performance in the portal. Figure 3.9 shows a view that highlights the overall resource utilization of a given query and its execution plan:

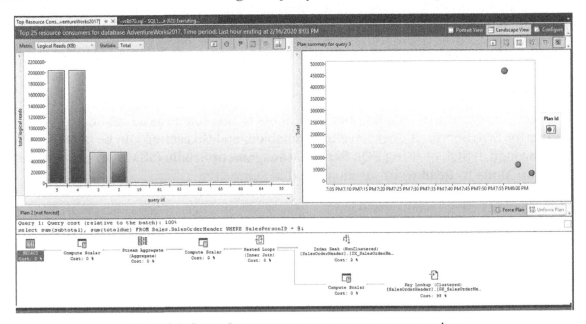

Figure 3.9: Query Store top resource consumers query view

In addition to being able to quickly identify poorly performing queries, or queries that have regressed in performance (a feature that can really help you mitigate any risks with an SQL Server upgrade), the Query Store allows you to force a given execution plan for a specific query. This can be useful when you have data skew and some parameter values for a query produce a poorly performing execution plan. This functionality is also built into the automated tuning feature that was introduced in SQL Server 2017; if a query's performance regresses, the database engine will revert to the last known good execution plan in an attempt to resolve the performance issue.

Azure portal

The Azure portal also provides a number of metrics to set a baseline for your VM workloads. By default, you will see the CPU average, the network bandwidth, the total disk bytes, and the disk operations per second (read and write) in the **Overview** blade for your VM. Data is available for the last 30 days, which is enough to set a solid baseline of your server's performance over time. This can help you easily find VMs that are over- or under-provisioned and help you track resources consumed.

In addition to this dashboard, you can allow metrics reporting to go beyond the performance counters in the **Overview** pane, as shown in *Figure 3.10*. This monitoring infrastructure can also connect to VM metrics, which provides alerting.

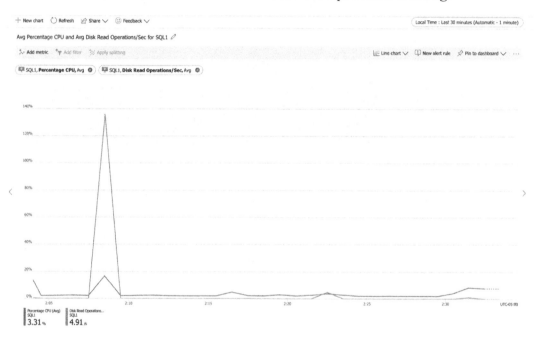

Figure 3.10: Azure VM metrics for an SQL Server VM

For example, if you wanted to report when a VM was using over 80% CPU over a period of 5 minutes, you would create a metric rule and then create an action group to be notified. You have a number of options as to what to do with the alert—you can do the standard SMS/email/push notification, or you can connect to a webhook that launches an action. You can also launch an Azure Automation runbook to carry out a remedial action within Azure. An example of where you might do this for SQL Server is to kick off an index report after a period of high CPU utilization.

Additionally, you collect and aggregate log and performance information from your Azure VMs using the Azure Diagnostics extension, which can connect to your Windows and application logs, and aggregate the logs into a number of destinations including Azure Monitor, Event Hubs, Azure Blob storage, and Application Insights. You can learn more about this functionality here[9].

Activity Monitor

The Query Store is one performance metric gathering option, but SSMS also has Activity Monitor, which provides an overview of all the activity on the server at any given time. It includes an overview of the processes running on the server, performance metrics, resource waits, and data file I/O. You can also customize columns to display more detailed information to meet your requirements. Below, in *Figure 3.11*, you can see Activity Monitor in operation:

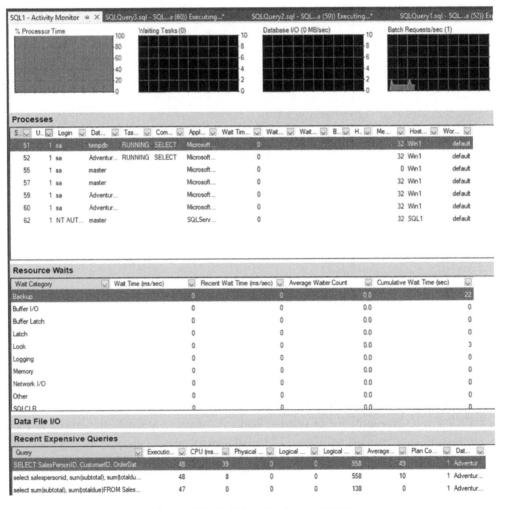

Figure 3.11 Activity Monitor in SSMS

Extended Events

SQL Server contains Extended Events, which is a lightweight performance monitoring system that can collect as much or as little information as needed to isolate a performance problem. Extended Events is structured into sessions, which are an event or group of events that have a target. The targets available for Extended Events sessions are:

- Event counter.
- Event file.
- Event pairing.
- Event Tracing for Windows.
- Histogram.
- Ring buffer.
- Azure Blob storage (Azure SQL Database only).

The other benefit of Extended Events is that you can use a predicate to limit the amount of data captured by the engine. SQL Server uses Extended Events sessions to monitor the health of Always On availability groups and general system health. You can also trace queries using Extended Events, including capturing actual execution plans. While this functionality can be helpful for gathering information, it should be carefully considered as there is a large amount of performance overhead associated with capturing actual execution plans.

Management Studio includes two Extended Events sessions under a header called xEvent Profiler, which is designed to emulate the functionality in SQL Server's Profiler tool. The xEvent Profiler tool provides a live view of queries streaming into the database server. This functionality is less intrusive and runs with less overhead than the older Profiler tool.

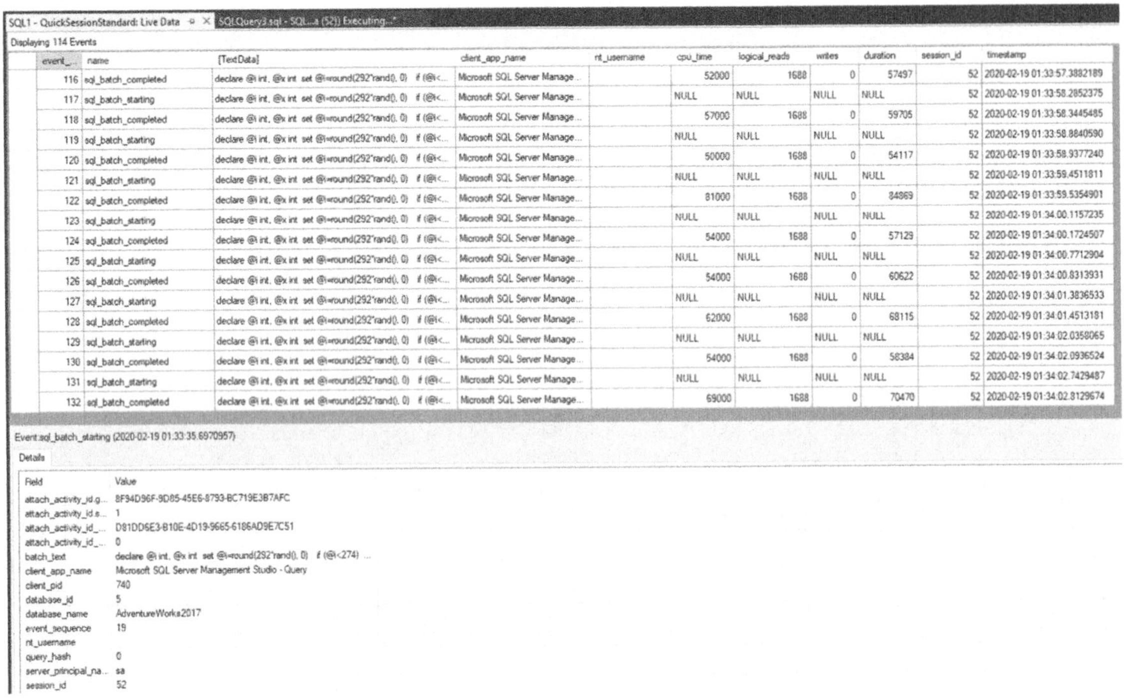

event_...	name	[TextData]	client_app_name	nt_username	cpu_time	logical_reads	writes	duration	session_id	timestamp
116	sql_batch_completed	declare @i int, @x int set @i=round(292*rand(), 0) if (@i<...	Microsoft SQL Server Manage...		52000	1688	0	57497	52	2020-02-19 01:33:57.3882189
117	sql_batch_starting	declare @i int, @x int set @i=round(292*rand(), 0) if (@i<...	Microsoft SQL Server Manage...		NULL	NULL	NULL	NULL	52	2020-02-19 01:33:58.2852375
118	sql_batch_completed	declare @i int, @x int set @i=round(292*rand(), 0) if (@i<...	Microsoft SQL Server Manage...		57000	1688	0	59705	52	2020-02-19 01:33:58.3445485
119	sql_batch_starting	declare @i int, @x int set @i=round(292*rand(), 0) if (@i<...	Microsoft SQL Server Manage...		NULL	NULL	NULL	NULL	52	2020-02-19 01:33:58.8840590
120	sql_batch_completed	declare @i int, @x int set @i=round(292*rand(), 0) if (@i<...	Microsoft SQL Server Manage...		50000	1688	0	54117	52	2020-02-19 01:33:58.9377240
121	sql_batch_starting	declare @i int, @x int set @i=round(292*rand(), 0) if (@i<...	Microsoft SQL Server Manage...		NULL	NULL	NULL	NULL	52	2020-02-19 01:33:59.4511811
122	sql_batch_completed	declare @i int, @x int set @i=round(292*rand(), 0) if (@i<...	Microsoft SQL Server Manage...		81000	1688	0	84869	52	2020-02-19 01:33:59.5354901
123	sql_batch_starting	declare @i int, @x int set @i=round(292*rand(), 0) if (@i<...	Microsoft SQL Server Manage...		NULL	NULL	NULL	NULL	52	2020-02-19 01:34:00.1157235
124	sql_batch_completed	declare @i int, @x int set @i=round(292*rand(), 0) if (@i<...	Microsoft SQL Server Manage...		54000	1688	0	57129	52	2020-02-19 01:34:00.1724507
125	sql_batch_starting	declare @i int, @x int set @i=round(292*rand(), 0) if (@i<...	Microsoft SQL Server Manage...		NULL	NULL	NULL	NULL	52	2020-02-19 01:34:00.7712904
126	sql_batch_completed	declare @i int, @x int set @i=round(292*rand(), 0) if (@i<...	Microsoft SQL Server Manage...		54000	1688	0	60622	52	2020-02-19 01:34:00.8313931
127	sql_batch_starting	declare @i int, @x int set @i=round(292*rand(), 0) if (@i<...	Microsoft SQL Server Manage...		NULL	NULL	NULL	NULL	52	2020-02-19 01:34:01.3836533
128	sql_batch_completed	declare @i int, @x int set @i=round(292*rand(), 0) if (@i<...	Microsoft SQL Server Manage...		62000	1688	0	68115	52	2020-02-19 01:34:01.4513181
129	sql_batch_starting	declare @i int, @x int set @i=round(292*rand(), 0) if (@i<...	Microsoft SQL Server Manage...		NULL	NULL	NULL	NULL	52	2020-02-19 01:34:02.0358065
130	sql_batch_completed	declare @i int, @x int set @i=round(292*rand(), 0) if (@i<...	Microsoft SQL Server Manage...		54000	1688	0	58384	52	2020-02-19 01:34:02.0936524
131	sql_batch_starting	declare @i int, @x int set @i=round(292*rand(), 0) if (@i<...	Microsoft SQL Server Manage...		NULL	NULL	NULL	NULL	52	2020-02-19 01:34:02.7429487
132	sql_batch_completed	declare @i int, @x int set @i=round(292*rand(), 0) if (@i<...	Microsoft SQL Server Manage...		69000	1688	0	70470	52	2020-02-19 01:34:02.8129674

Field	Value
attach_activity_id.g...	8F94D96F-9D85-45E6-8793-BC719E3B7AFC
attach_activity_id.s...	1
attach_activity_id_...	D81DD6E3-B10E-4D19-9665-6186AD9E7C51
attach_activity_id_...	0
batch_text	declare @i int, @x int set @i=round(292*rand(), 0) if (@i<274) ...
client_app_name	Microsoft SQL Server Management Studio - Query
client_pid	740
database_id	5
database_name	AdventureWorks2017
event_sequence	19
nt_username	
query_hash	0
server_principal_na...	sa
session_id	52

Figure 3.12: Azure xEvent Profiler in SSMS

Extended Events is a very deep topic and covers almost all of the functionality in the database engine. You can learn more about all of the events that are available here[10].

Identifying disk performance issues with SQL Server

As mentioned earlier in the chapter, disk performance is critical to database systems, and this can be exaggerated in an Azure environment where latencies may be slightly higher (for deployments other than ultra disk deployments) than in a high-performance on-premises environment. There are a couple of different ways to measure I/O performance on an SQL Server. The first and most common way is to query the `sys.dm_io_virtual_file_stats` dynamic management view. You can also query `sys.dm_os_wait_stats` to identify what the server is waiting for. High percentages of `pageiolatch_xx` waits can be indicative of storage issues. You can also validate the data reported by SQL Server by capturing data from the Linux **iostat** command to report on the performance of the devices. On Windows Server deployments of SQL Server, you can use the built-in performance monitor (perfmon) capabilities to capture performance data. This post[11] on Microsoft Docs offers more detail on how to identify and troubleshoot a performance issue.

Key performance features in SQL Server

In addition to the monitoring and metric capabilities that both SQL Server and Azure provide, SQL Server provides many features that make use of in-memory technology set to deliver world-class performance:

- **In-Memory OLTP tables**: An in-memory latchless data structure that delivers extremely fast insert performance.

- **Hybrid Transactional/Analytical Processing (HTAP)**: This technique combines filtered non-clustered columnstore indexes with in-memory OLTP tables to deliver fast transaction processing and to concurrently run analytics queries on the same data.

- **TempDB**: The memory-optimized TempDB metadata feature effectively removes some contention bottlenecks and unlocks a new level of scalability for TempDB-heavy workloads.

These features are typically implemented in conjunction with new application development. You can learn more about these features here[12].

Security concepts

Azure offers a number of built-in and optional security features that help you build a more secure environment. There are a number of options, including network security groups, disk encryption, and key management, that help you ensure your security. Azure is the most compliant database for your SQL installations. You can read more about this here[13].

Let's examine the specific security features Azure provides:

- **Azure Security Center (ASC)** is your centralized security management system in Azure that provides advanced threat protection for your hybrid workloads in the cloud. Using ASC, you can configure security policies for your VMs, detect threats to your VMs and SQL databases via real-time alerts, and mitigate them using ASC's recommendations.

- Advanced data security for SQL Server on Azure VMs is another Azure-specific security feature. It integrates with ASC and enables the detection and mitigation of potential database vulnerabilities and threats.

- Key management for encryption is enhanced in Azure using Azure Key Vault (**AKV**), which enables you to bring your own key and store it in AKV to manage the encryption and decryption of your databases.

- In addition to the above capabilities, SQL Server Azure VMs can use automated patching[14] to schedule the installation of important Windows and SQL Server security updates automatically.

Connecting to Azure VMs

It is important to note that, by default, in the Azure portal, new VMs are created with a public IP address. This is something you should not do for any SQL Server with production data, but if you require a public IP address, you should limit the connections to only the IP addresses that should connect to the VM. When creating this VM, you also have the option to open ports such as 1433 to the internet.

> **Note**
>
> Opening an SQL Server to the public internet will result in a number of failed logins from botnets across the world and may impact the performance of the server. It is something that you should pretty much never do.

Fortunately, Azure provides a number of other ways to connect to your VMs. If you have a site-to-site VPN or Express Route connection from your on-premises network into your virtual network, you can connect to your VM just as if it was in your on-premises datacenter. Azure is just an extension of your datacenter and network once you have a VPN connection. Once you have a VPN connection in place, connecting to a VM in Azure is no different than connecting to a server in a different datacenter. Another option is using Azure Bastion, which is a service that allows a secure connection over port 443 from the Azure portal into your VM using a desktop emulator.

Network security groups

Azure Virtual Networks can be split into subnets. These subnets allow you to isolate your network traffic between various application tiers, as shown in *Figure 3.13*. **Network security groups (NSGs)** act as firewalls between these subnets. NSGs contain security rules that filter network traffic inbound to and outbound from a virtual network subnet by IP address, port, and protocol. These security rules are applied to resources deployed within the subnet.

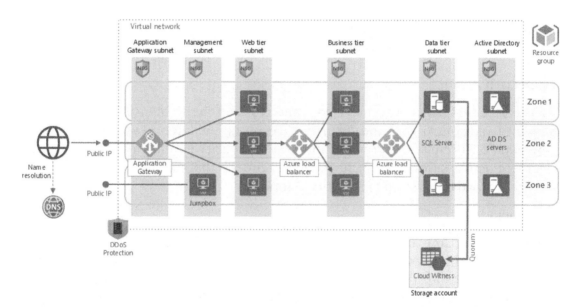

Figure 3.13: Complex Azure network architecture

As we can see in *Figure 3.13*, there is a single virtual network divided into six subnets. Without going into the full detail of the architecture, there is a public IP address that points to an application gateway, which allows traffic on port 443, and performs SSL termination. The web tier then points to a business subnet, where traffic is directed to a load balancer over the port used by the app server. Finally, the data tier allows traffic only from the business tier into port 1433, and the SQL Server instances have secure access into a storage account for backups and Cloud Witness. While an NSG can be applied on the virtual network, subnet, or even individual virtual NIC, they are most commonly deployed at the subnet level.

> **Note**
>
> Azure Firewall offers some more functionality than NSGs and may be required for some deployments. You can learn more about the differences between the services here[15].

In the age of ransomware, proper network segmentation is critical to the security of your data. One of the major benefits of Azure is that you can easily segment your network without any cabling changes.

Azure Security Center

ASC is a service that helps prevent, detect, and respond to threats with increased visibility into and control over the security of resources and hybrid workloads in the cloud and on-premises. It provides integrated security monitoring and policy management across subscriptions, helps detect threats that might go unnoticed, and works with a broad ecosystem of security solutions.

Features include:

- **Threat protection**: ASC's threat protection includes fusion kill-chain analysis, which automatically correlates alerts based on cyber kill-chain analysis to better understand attack campaigns, providing details on where they started and the impact they had on resources. Other capabilities include the automatic classification of data in Azure SQL, assessments for potential vulnerabilities, and recommendations for how to mitigate them.

- **Secure Score**: Secure Score is a feature that reviews security recommendations provided by ASC and prioritizes them, targeting the most serious vulnerabilities for investigation first.

- **Azure Policy**: ASC allows enterprises to define their specific security requirements and configure workloads through Azure Policy. Recommendations will then be based on those policies and can be customized as needed.

- **Azure Monitor**: Azure Monitor maximizes the availability and performance of applications and services by delivering a comprehensive solution for collecting, analyzing, and acting on telemetry from cloud and on-premises environments. As a security tool, it helps control how data—including sensitive information such as IP addresses or user names—is accessed.

- **Security Posture**: ASC uses monitoring capabilities to analyze overall security and identify potential vulnerabilities. Information on network configuration is available instantly.

Authentication

SQL Server on both Windows and Linux offer both SQL Server and Windows Authentication (Active Directory). Active Directory authentication allows users to log in using a single sign-on without being prompted for a password. This authentication is provided by using Kerberos based on the connection with the Active Directory domain controllers. In Windows, this feature is provided by joining your server to your Active Directory domain. In Linux, the process is slightly more complicated.

To configure Windows Authentication on SQL Server on Linux, you need at least one domain controller; you have the option of using the **realmd** and **sssd** packages on your Linux VM in order to join the VM to your Windows domain. This is the preferred method, but you also have the option to use partner LDAP utilities. You can follow the instructions in this <u>documentation</u>[16] to configure your VM for Active Directory authentication.

SQL Server authentication is configured by default on SQL Server on Linux installations. While SQL Server authentication is easier to configure, it has a few disadvantages compared to Windows authentication:

- There is no built-in syncing of logins across servers, and users have to remember additional passwords.

- SQL Server cannot support the Kerberos protocol.

- Windows authentication offers additional security and password policies that SQL Server authentication does not support.

- The encrypted password is passed over the network at the time of authentication, making it another point of vulnerability.

SQL authentication does allow for a wider variety of SQL clients and can be required by some older applications.

SQL Server security

Beyond authentication, SQL Server provides a robust set of permissions and privileges to manage security at each layer—server, database, object, and all the way down to columns. The database engine includes a set of built-in roles and allows logins and users to be defined at the server and database level. Alternatively, users can be scoped to a specific database using contained user functionality. This is typically used for applications that only connect to a single database. You can learn more about the security features of SQL Server here[17].

Advanced data security for SQL Server on Azure VMs

Advanced data security for SQL Server on Azure VMs is a new security feature that includes Vulnerability Assessment and Advanced Threat Protection. This feature includes functionality for identifying and mitigating potential database vulnerabilities and detecting anomalous activities that could indicate threats to your database. Some of the tools included to perform these tasks are detailed below:

- Vulnerability Assessment[18] is an easy-to-configure service that gives you visibility into your configuration, databases, and data. The tool runs a scan on your database using a knowledge base of best practices and looks for excessive permissions, unprotected sensitive data, and misconfigurations. The assessment products a report that tells you how to remediate the issues that the report finds.

- Advanced Threat Protection[19] detects anomalous activities indicating unusual and potentially harmful attempts to access or exploit your SQL Server. It continuously monitors your database for suspicious activities and provides action-oriented security alerts on anomalous database access patterns. These alerts provide the details of any suspicious activity and recommended actions to investigate and mitigate threats.

- Integration with ASC provides benefits such as email notifications for security alerts, with direct links to alert details. You can also explore Vulnerability Assessment reports across all your databases, along with a summary of passing and failing databases, and a summary of failing checks according to risk distribution. Also, you can use ASC to explore and investigate security alerts and get detailed remediation steps and investigation information in each one.

Azure Active Directory

Azure Active Directory is not supported in SQL Server (it is supported for Azure SQL Database and Azure SQL Database managed instances). However, it can play a key role in some VM automation scenarios. You can create a managed service identity for your VM—this is somewhat similar to the concept of a service account in Windows.

The following figure illustrates the Azure managed identity configuration process:

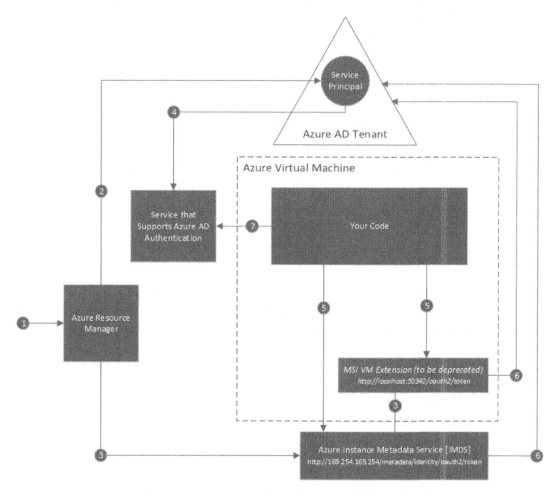

Figure 3.14: Azure managed identity configuration

You can use this identity with a number of Azure services, including Azure Storage and Key Vault, with the identity of the VM. This can allow you to securely execute scripts within your VM using the identity of the VM to authenticate itself.

Azure Key Vault

AKV is a security service that helps with management tasks, key access, and certificate management—all of which are protected by hardware security modules that meet FIPS 140-2 Level 2 requirements. For the sake of simplicity, AKV can be considered as Azure's password manager. Beyond simply being a password manager, AKV enables granular access of secrets by applications, services, and monitoring of its own use.

Where AKV integrates with SQL Server is that you have the option to store your Always Encrypted certificates in a key vault. Additionally, you can store any certificates (for example, **transparent data encryption (TDE)**, backup certificates, or client connection certificates) that are used in your SQL Server environment in a key vault.

SQL Server provides a connector for AKV that allows AKV to serve as an Extensible Key Management solution for storing SQL Server certificates and keys. This provides a critical backup in case any of those certificates are lost, in addition to securing their storage. You can learn more about this process here[20].

Transparent data encryption

SQL Server provides different types of encryption. The first and most basic is TDE, which provides encryption at rest for data files. This protects you from an attacker gaining access to the VM's data drive and taking either backups or data files. The following figure illustrates the TDE architecture:

Transparent Database Encryption Architecture

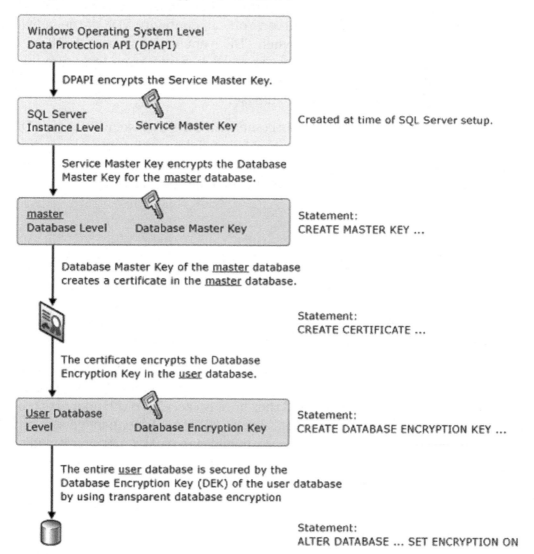

Figure 3.15: TDE architecture

This meets the common requirement of encryption at rest, and also encrypts backups by default. This feature was an Enterprise Edition feature until SQL Server 2019, when it became available in all editions of SQL Server. You can configure TDE at the individual database level, but you should note that when TDE is enabled for a user database, it is enabled on the TempDB system database as well. It is important to ensure the certificates you used when configuring TDE are backed up—without them, you will not be able to restore a database or attach a database.

In addition to these security features, VMs come with their own security, Azure Disk Encryption—a feature that helps protect and safeguard data and meet organization and compliance commitments. So, in the case of TDE customers, they get multiple encryption protections—Azure Disk Encryption and encryption through the SQL Database host.

Always Encrypted

While TDE is designed to meet the requirements of encryption at rest, administrators and users who have access to the database have full access to the unencrypted data, where it can be consumed and potentially exported in a tool such as Excel. Additionally, with any other encryption solution, such as SQL Server's cell-level encryption, database administrators have access to the encryption keys.

Always Encrypted changes this paradigm—the encryption key for the encrypted data is accessed in the client application and never in the database server. The administrators have no access to the encryption key, and therefore no access to the unencrypted values. Having a key management process, such as AKV, enables enhanced separation of duties to prevent this administrator access. In addition to AKV, options include Windows Certificate Store on a client machine, or a hardware security module.

> **Note**
>
> You should never generate the keys for your Always Encrypted columns on the server hosting your database, as anyone with access to the server could potentially gain access to keys in memory.

Always Encrypted is designed to protect sensitive data. It should be applied as an additional layer in tandem with TDE and TLS capabilities and should be used very selectively to protect data such as national ID or U.S. Social Security numbers, or other sensitive fields, and not broadly across all of your tables and columns. You can see more examples of how to use the feature here[21].

Since the application client has access to the keys to the VM, if the application executes the following query, the query will be sent to the database server with the SSN value in encrypted ciphertext, where it will be executed against the ciphertext in the **Patients** table:

```
SELECT FirstName, LastName FROM Patients WHERE SSN='111-22-3333'
```

This is shown in *Figure 3.13*:

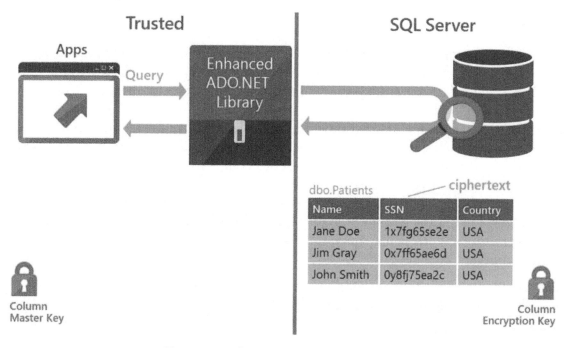

Figure 3.16: Always Encrypted architecture

The Always Encrypted architecture shows that a column master key is created for the application, and a column encryption key is stored in the database. Always Encrypted offers two kinds of encryption, deterministic and randomized. Deterministic encryption means the same value will have the same ciphertext every time—this allows **GROUP BY**, equality joins, and indexing, among other benefits. While this may be acceptable for many scenarios, columns that have a small set of possible values can be identified by guessing. Randomized encryption should be used where data is not grouped with other data and not where it is used to join tables.

SQL Server 2019 adds even more functionality to the Always Encrypted feature set by adding secure enclaves functionality, which allows the database engine to operate on encrypted data. In earlier versions of Always Encrypted, SQL Server could not perform computations or pattern matching on the data. An enclave is a protected memory region within the SQL Server process that can access encrypted data. The keys are never shown in the engine plaintext during query processing. You can learn more about secure enclaves here[22].

Dynamic data masking

While Always Encrypted protects data via encryption, and data is never stored in an unencrypted fashion, dynamic data masking does not touch the underlying data in the database. Dynamic data masking is implemented in the presentation layer, which means it is very easy to implement in an existing application with minimal or no changes to the application code. One interesting use case of dynamic data masking is to randomize the data in sensitive columns, and then export the randomized data to non-production environments (the feature should be considered complementary to other security features, such as Always Encrypted, row-level security, and auditing. Also, note that administrators always have access to the unmasked data). However, it can be a very good feature to limit the amount of data users can see in an application. For example, call center representatives may need to identify customers by the last four digits of an account number.

Dynamic data masking is implemented at the presentation layer, which means it is very easy to implement in an existing application with minimal or no changes to the application code. One interesting use case of dynamic data masking is to randomize the data in sensitive columns, and then export the randomized data to non-production environments.

Azure Disk Encryption

In addition to these security features, VMs come with their own security. In the case of TDE, customers get multiple layers of encryption: Azure Disk Encryption provides encryption at rest for the operating system and data disks associated with the VM, and SQL Server Database also encrypts the data and log files, as well as the backups inside of the operating system. Azure Disk Encryption uses BitLocker on Windows and DM-Crypt on Linux VMs.

You can learn more about Azure Disk Encryption here[23].

Auditing

SQL Server provides two types of auditing–server and database auditing. Server audits capture instance-level events, such as backups and restores, database creation and removal, logins, and numerous other options. You can see the full list of server audit events <u>here</u>[24]. Database-level auditing allows you to track query execution by users and can be configured on specific objects in the schema.

Data Discovery and Classification

One of the recent additions to SQL Server is the Data Discovery and Classification feature. This was first introduced into SSMS using extended properties on database objects. Starting with SQL Server 2019, this functionality is built into the database engine. SQL Server will identify columns (based on names and pattern matching) that could potentially contain sensitive data, which provides an easy way to review and appropriately classify your data. You can then label the columns using tags, which provides visibility in reporting for both compliance and audit purposes. SQL Server 2019 stores this data in a catalog view called `sys.sensitvity_classifcations`.

Summary

One of the benefits of building an SQL Server VM on Azure is that you can quickly get started learning about the features of SQL Server while configuring a limited amount of infrastructure. Azure makes it possible to simulate complex network architecture with a few lines of code, which allows you to protect your data. SQL Server and Azure also offer a wide variety of performance features that allow you to understand the workloads on your system and troubleshoot problematic queries and workloads. In addition, SQL Server provides best-in-class security features to meet the most stringent encryption needs. In fact, newer security offerings such as Advanced Data Security are only available for SQL Server instances on Azure VMs and not on-premises as yet–an advantage over traditional SQL Server on-premises installation. In the next chapter, you will learn more details about running SQL Server on Linux VMs in Azure.

Chapter links

1. https://bit.ly/2X0VExf
2. https://bit.ly/2WXwu2G
3. https://bit.ly/36rHusa
4. https://bit.ly/3c2Z2vX
5. https://bit.ly/3cZDIss
6. https://bit.ly/2LYy8dW
7. https://bit.ly/2WZoLkt
8. https://bit.ly/2XsY2Mj
9. https://bit.ly/3bZthDZ
10. https://bit.ly/3grhUID
11. https://bit.ly/3gnwUXA
12. https://bit.ly/36tJ9gR
13. https://bit.ly/3efwJf7
14. https://bit.ly/3c30D4V
15. https://bit.ly/3gkJilc
16. https://bit.ly/3griKVN
17. https://bit.ly/3ddGxGJ
18. https://bit.ly/2X2HdsO
19. https://bit.ly/2B2Urxb
20. https://bit.ly/3c0L8ub
21. https://bit.ly/2XnXMhE
22. https://bit.ly/3gnZp7B
23. https://bit.ly/2AXajkx
24. https://bit.ly/2TCbFYo

SQL Server on Linux in Azure Virtual Machines

By Anthony Nocentino

In *Chapter 3, Hero capabilities of SQL Server on Azure VMs*, we learned about some of the capabilities and features of running SQL Server on Azure Virtual Machines. In this chapter, let's shift the conversation toward leveraging SQL Server on Linux to build your next application in Azure. We will look at the development ecosystem around SQL Server on Linux and how SQL Server can help add value to your business applications. We'll introduce the languages, frameworks, and tools available for you to build and manage your data estate in Azure. Once we have a firm understanding of how we can develop applications and the tools available to work with our Azure-based resources, we will focus on the platforms available for running SQL Server in Azure and helping you decide what is best for your applications. We will also look at the Azure services available to help you build scalable and maintainable SQL Server-based applications in the cloud.

Let's begin by taking a look at open-source languages, frameworks, and tools available to work with SQL Server on Linux.

SQL Server on the Linux development ecosystem

Modern SQL Server provide an open-source developer with the ability to opt for the languages and tooling of their choice in order to successfully build their application. In this section, let's take a look at the language choices you have when developing code and the tools available when working with SQL Server.

Open-source development frameworks and tooling for SQL Server on Linux in Azure Virtual Machines

For years, developers had to choose between open-source development frameworks and those frameworks available for Microsoft technologies. Well, over the course of the last few years, that binary decision has changed dramatically. Microsoft has made a conscious effort to enable and support many different developer frameworks and technologies, embracing open-source technologies within the Microsoft platform. Now available to you are connection modules (drivers) for the programming languages of your choice, such as *Ruby*, *PHP*, *R*, *Python*, and *Java*. For access to these connection modules and more information on how to use them, check out this documentation[1]. Let's take a quick tour of the languages, the extensibility framework, and language extensions available to help you develop applications using SQL Server. The tools and techniques mentioned in this section support both SQL Server on Linux and also SQL Server on Windows. Therefore, code and tools developed can be targeted toward either platform.

The following client libraries are available for interacting with Microsoft SQL Server:

- **Java**
- **Node.js**
- **PHP**
- **Python**
- **R**
- **Ruby**
- **C#**
- **C++**

If you're just getting started in the Microsoft space and want to see what it takes to build an application with your preferred development language on your preferred platform using SQL Server as your data store, check out this link[2] for more information. Here, you'll be presented with the instructions to build an application right away. We highly recommend starting here to see what's available to you now.

Knowing the collection of languages and client libraries available to developers, now let's look at how SQL Server technologies can be leveraged to increase business values and functionality in applications.

The extensibility framework and language extensions

As a developer, you are used to the paradigm of writing code that retrieves data from a relational database and then processes that data inside your application. What if we told you that you could write application code and store it, execute it, and process data inside SQL Server and return the results back to your client application? With SQL Server's extensibility framework[3], and language extensions[4], you can work with complex data analysis and machine learning scenarios in your application architecture. This paradigm shift in development techniques provides two benefits—you can use tools of your choice for your projects and also boost performance since the computing and data processing is occurring inside SQL Server. Let's look at both more closely.

First, under the SQL Server machine learning services[5] umbrella in SQL Server, you can leverage the tools, packages, and models you're used to using as a developer or data scientist in both R[6] and Python[7].

With the language extensions feature of SQL Server, you'll find that you have the ability to run Java code inside the database engine itself. With language extensions, your code is stored and executed inside SQL Server, providing benefits such as security and performance. In terms of security, you can place access controls around the execution of your application code. Furthermore, since the execution of your code is inside the database engine, your data does not have to traverse potentially insecure networks. Avoiding this network transport also provides a performance benefit since all of the processing and data access takes place locally inside the database engine.

Language extensions are a subset of the extensibility framework in SQL Server. Language extensions provide developers with the ability to run external code inside SQL Server. In SQL Server 2019, Java is available as a language extension, enabling you to write classes that can execute inside SQL Server and directly access set-based data. This provides performance and security benefits. For more information on how to use Java programs inside SQL Server, check out this Microsoft documentation[8].

In addition to the languages available as part of SQL Server's extensibility framework, developers can use familiar object-relational mapping frameworks to enable rapid application developments.

Object-relational mapping (ORM) frameworks

The final stop on our tour of languages and frameworks supported by SQL Server is object-relational mapping (ORM) frameworks. ORMs enable developers to write code as objects and allow the database connectivity driver to convert that runtime object into structured data stored in the relational database, a valuable tool for rapid application development. SQL Server is supported as a back-end data store to several popular ORM frameworks. Some of these are listed here:

- Entity Framework and Entity Framework Core[9]
- Hibernate[10]
- Laravel Eloquent[11]
- Sequelize[12]
- Django[13]
- Ruby on Rails[14]

So now that we have an idea of what languages and frameworks are available to us for developing on SQL Server, let's move on and learn about the tools available for developing our applications and managing our platforms in Azure.

Cross-platform tooling

Microsoft offers a traditional suite of tools[15] to manage both SQL Server and Azure and recently, many of these tools have been made available on multiple platforms. Previously, developers using Linux or Mac workstations would need to run a Windows VM to develop applications in Visual Studio and to manage SQL Servers with SQL Server Management Studio, but this is no longer necessary. Available now are cross-platform tools such as Visual Studio Code to develop applications and Azure Data Studio to manage a SQL Server estate and develop T-SQL directly without the need for a Windows-based VM. Developers and administrators can use these cross-platform tools to develop and manage applications and systems across their entire SQL Server estate, including servers on both Windows and Linux platforms.

Let's have a look at the different sets of tools available for application and database development:

Graphical tools

- **Visual Studio Code**: A graphical IDE that provides syntax highlighting, debugging, and Git integration. Visual Studio Code[16] extensions provide a wide library of additional languages, debuggers, and tools to enable rapid and efficient development.

- **Azure Data Studio**: Based on a VS Code core, Azure Data Studio[17] provides similar features and functionality, but with a focus on data and managing SQL Server estates. It has additional user experiences geared toward working with data both on-premises and in the cloud.

An interesting note regarding both Visual Studio Code and Azure Data Studio is the fact that both of these tools are being developed out in the open on GitHub and accept pull requests from the community! If there is a feature you would like to be added or a bug you know how to fix, then go ahead and check out Microsoft's azuredatastudio[18] and vscode[19] repositories respectively to start participating in projects.

On a personal note, I have been using the Mac and Linux platforms for nearly 20 years. I have had to keep a VM around to manage systems and develop applications running on the Microsoft platform. Tools such as Visual Studio Code and platform choices such as running SQL Server on Linux have directly impacted how I consume these products and develop solutions. The main idea here is choice—I am able to choose what platform I want to use when working with these technologies.

In addition to the aforementioned graphical tools, many cross-platform, command-line tools are now available. Each of these tools is used for various development and management scenarios:

Command-line tools

- **PowerShell Core**: A command-line scripting language used for automation tasks. PowerShell, familiar to Windows administrators for many years, is now a cross-platform tool that can be used on nearly any base operating system to manage systems and applications and also build integrations and pipelines. Check out this Microsoft documentation[20] for installing PowerShell.

- **mssql-cli**: Provides an interactive command-line interface experience for querying with SQL Server. It's a new tool[21] available for developers and DBAs to interact with SQL Server interactively at the command line with autocompletion. This enables you to quickly browse and work with data available from SQL Server.

- **sqlcmd**: A command-line utility for running T-SQL on SQL Server. This tool is familiar to many DBAs and developers for interacting with SQL Server at the command line. This tool[22] is now available on Windows, Linux, and Mac, enabling cross-platform management of your SQL Server.

- **Cloud Shell**: Provides access to all of your Azure command-line tools available in a browser window. The value that Cloud Shell adds is that the tools in Cloud Shell will be updated and managed by Microsoft. So, as new versions of the tooling become available, you do not have to maintain Cloud Shell—it is automatically updated for you. Tools such as Bash, PowerShell, SQL Server command-line tools, and the Azure CLI are all available in Cloud Shell[23].

Now, transitioning from development into production, let's move on from the languages and tools available to build applications and discuss the platforms and tools available that are used to deploy SQL Server and SQL Server-based applications in Azure.

Platform deployment and management for SQL Server in Azure

DevOps and infrastructure as code have been trending in IT development and operations processes for the last several years, enabling developers and operations teams to write code to define the desired state of the infrastructure needed to support applications. Infrastructure as code-based solutions enable developers and operations teams to build repeatable and tested solutions. To enable these processes and scenarios, Microsoft provides the Azure CLI[24] and Azure PowerShell[25] (the PowerShell Az module). These tools are the foundation for any programmatic interaction with Azure and are available on Windows, Linux, Mac, and Docker.

You can build custom solutions with both the Azure CLI and Azure PowerShell at the command line using imperative techniques, where you execute the commands necessary to build the Azure infrastructure and resources requested. However, to really enable infrastructure as code, you will need to use a declarative technique to write documents that describe the Azure infrastructure and resources that need to be built. The ARM templates[26] enable you to build configuration documents that describe your Azure infrastructure and its resources and build repeatable, declarative processes for deployment. This code can be managed in source control and used as part of a deployment pipeline to ensure that your Azure infrastructure and its resources are in the desired state prior to rolling out applications.

In addition to the command-line tools available to manage your Azure infrastructure and resources, the Azure portal has a graphical, interactive way to create and manage resources in Azure. The Azure portal can be used to create individual resources with predefined workflows, but can also be used to deploy Azure ARM templates, which can be uploaded to the Azure portal interactively.

We encourage you to create a Linux SQL Server VM in the Azure portal. For step-by-step instructions on how to do so, check out this Microsoft documentation[27]. One of the main benefits of creating a VM in the Azure portal is finding what's new. The Azure portal will surface new configurations and options graphically so that you can easily see what's changed when deploying a resource, in this case, our VM.

When using the Azure portal to create resources, one of the final options available to you when completing the wizard to create a resource is to **Download a template for automation**. While you can still create the resource in the portal by clicking on the **Create** option, the download option will give you the ARM template that you can use to source your automation tasks. This process of exporting your ARM template and using it as a source for your automation tasks enables you to quickly generate your infrastructure as code resource configuration document (ARM template) and use that document as part of your deployment pipelines when combined with the Azure CLI or Azure PowerShell.

The tools to help deploy and manage resources available in Azure are core to building fast and repeatable deployments. Now, let's drill down further and talk about the platform options available for running SQL Server on Linux in Azure IaaS VMs.

Supported base operating systems for running SQL Server on Linux in Azure IaaS VMs

When deploying SQL Server on Linux in Azure, you have several platform choices. In this section, I will introduce how to deploy SQL Server on Linux in Azure using an Azure Marketplace image with SQL Server on Linux pre-installed, using an Azure Marketplace image and installing SQL Server on Linux manually, and also container-based deployments. I will wrap up this section with advice on which platform you should choose for your applications.

Using an Azure Marketplace image with SQL Server on Linux pre-installed

When creating Azure VMs to run SQL Server on Linux, the underlying platform becomes a critical choice in terms of deployment and management.

When it comes to deploying SQL Server on Linux in Azure, you can start off by using pre-built Azure Marketplace images. This means that you can select a base operating system with SQL Server on Linux pre-installed. There are Marketplace images with SQL Server pre-installed available for the latest versions of currently supported Linux distributions. You can also choose between versions of SQL Server, including both 2017 and 2019, and editions such as Developer, Standard, or Enterprise. In *Figure 4.1*, you can see the Azure Marketplace images available for running SQL 2019 on Ubuntu or Red Hat Enterprise Linux:

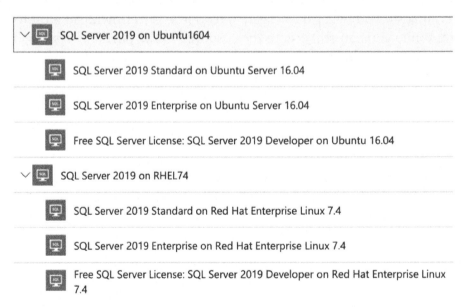

Figure 4.1: Selecting an Azure Marketplace image for SQL Server on Linux 2019

You can find more information on working with a **SQL IaaS VM** in Azure here[28].

Using a Linux Azure Marketplace image and installing SQL Server on Linux manually

Now, there are scenarios where you'll need to deviate away from these Azure Marketplace images with SQL Server on Linux pre-installed. Perhaps you need to run a supported Linux distribution or version that is not currently available as a Marketplace image, for example, Red Hat Enterprise Linux 8. If that is the case, you can deploy the base operating system image from the Azure Marketplace and use that operating system package manager to install SQL Server on Linux:

- For more information on installing SQL Server on Linux via a package manager, check out this Microsoft documentation[29].

- For more information on currently supported operating systems and platforms for running SQL Server on Linux, check out the Red Hat Enterprise Linux[30], SUSE Linux Enterprise Server[31], Ubuntu[32], and Docker[33] quickstart and installation guides.

- For information on configuring SQL Server on Linux for high availability using availability groups on IaaS VM deployments in Azure, refer to this documentation[34].

Container-based deployments for SQL Server on Linux in Azure

When SQL Server became available on Linux, it opened the door for SQL Server to run in containers. A container is a form of operating system virtualization that allows you to run multiple applications on the same base operating system without the applications and their processes knowing about each other. This isolation is central to the success of containers as it means that scenarios in which applications may conflict with one another can now be deployed on the same system. A container is a running container image. A container image contains the binaries, libraries, and filesystem components to run your application. Let's take a second to explore some of the benefits of running applications in containers:

- **Speed**: When compared with VMs, containers are significantly smaller, meaning they are more agile in moving around inside a datacenter. A VM running SQL Server consumes about 60 GB+ of storage before any databases are created—just the base operating system plus the SQL Server installation. A container image for SQL Server is about 1.5 GB. Copying 1.5 GB of data is straightforward for any modern datacenter or even a home internet connection.

- **Upgrades**: We can upgrade SQL Server by creating a new container running the newer version of SQL Server. Moreover, within major versions, you have the ability to roll back if needed.

- **Fast and consistent deployments**: Running SQL Server in containers enables us to write code for our deployments and common deployment mechanisms to roll out SQL Server in a fast and consistent way enabling DevOps practices.

- **DevOps**: Running SQL Server in containers, you can leverage SQL Server as an element of your Continuous Integration and Continuous Deployment pipelines.

- **Testing**: Containers give SQL Server users the ability to quickly create a SQL Server instance and consume its services for testing scenarios, perhaps with a view to validating version upgrades or patches.

Running multiple SQL Server containers on an IaaS VM in Azure

VMs are the dominant compute layer for running application containers in the cloud. By deploying a VM with a supported base operating system, such as Ubuntu or Red Hat Linux, you can quickly deploy multiple containers running SQL Server onto the base operating system, thereby increasing the density of your SQL Server deployment footprint on that single VM. This deployment paradigm is similar to using named instances[35] in Windows-based deployments of SQL Server, where you can have multiple, unique SQL Server instances running on a single base operating system. In container-based deployments, each container started on the base operating system has a unique instance of SQL Server and is available for the application to connect to over the network available on its own unique TCP port. The isolation provided by containers enables you to confidently run multiple versions and editions of SQL Server on the same base operating system when deploying SQL Server in multi-container scenarios.

Container images available for SQL Server on Linux

SQL Server on Linux is available in several container image permutations of SQL Server versions and container operating systems. You will find SQL Server container images for SQL Server 2017 and 2019 that are based on Red Hat Enterprise Linux and Ubuntu. Since there are many combinations available, it's important to be able to quickly find which container images are available, along with their platform and version. This plethora of numerous, readily available images enables developers to be able to quickly access, deploy, and test applications against the many versions and iterations of SQL Server available.

The following code shows how to generate a list of container images available for Red Hat Enterprise Linux (output abbreviated):

```
curl -sL https://mcr.microsoft.com/v2/mssql/rhel/server/tags/list
```

```
"2019-CU1-rhel-7.6"
"2019-CU1-rhel-8"
"2019-GA-rhel-7.0"
"2019-GA-rhel-7.6"
"2019-GDR1-rhel-7.0"
"2019-GDR1-rhel-7.6
"2019-latest"
"latest"
"vNext-CTP2.0"
```

The following code shows how to generate a list of available container images available for Ubuntu Linux (output abbreviated):

```
curl -sL https://mcr.microsoft.com/v2/mssql/server/tags/list
```

```
"2019-CTP3.2-ubuntu"
"2019-CU1-ubuntu-16.04"
"2019-GA-ubuntu-16.04"
"2019-GDR1-ubuntu-16.04"
"2019-RC1"
"2019-RC1-ubuntu"
"2019-latest"
"latest"
"latest-ubuntu"
```

Having selected a container platform, let's move on to how to run SQL Server in a container.

Starting a container running SQL Server on Linux

The simplicity and speed of starting containers is quite remarkable. The following code will start a container running SQL Server 2019 CU1 on an Ubuntu Linux-based container image. Run this code and seconds later you will have a running instance of SQL Server to connect applications to. A word of caution—this container definition is NOT using persistent storage and, if deleted, any data inside this container will be deleted:

```
docker run \
    --env 'ACCEPT_EULA=Y' \
    --env 'MSSQL_SA_PASSWORD=S0methingS@Str0ng!' \
    --name 'sql1' \
    --publish 1433:1433 \
    --detach
    mcr.microsoft.com/mssql/server:2019-CU1-ubuntu-16.04

 94936a1b517650426df8fdc896fa4ceac3fb553326f52575a384fe030e8f04de
```

Here are a few critical scenarios deploying SQL Server on containers, including how to persist data when using containers:

- Running multiple SQL Server containers

- Persisting your data

- Upgrading SQL Server in containers

- Troubleshooting

- Building a custom SQL Server on Linux containers

To read about these in detail, refer to this Microsoft documentation[36].

Deploying SQL Server in containers in Azure

In addition to deploying SQL Server in containers on IaaS VMs, as introduced in the previous section, *Container-based deployments for SQL Server on Linux in Azure*, you can use Azure Container Instances and Azure Kubernetes Service to deploy SQL Server in Azure:

- Azure Container Instances[37]: For serverless implementations of containers in Azure, commonly used to stand up containers to support short-term workloads.

- Azure Kubernetes Service[38]: For production-grade, container-based workloads. AKS is a platform commonly used for long-running, container-based workloads in Azure.

If you look at the most recent product releases from the SQL Server product group, you will find solutions such as Big Data Clusters and Azure Arc data services. These solutions are built entirely on containers running in container orchestration platforms, both in on-premises and cloud scenarios. The future of SQL Server is containers. For me, a few years ago, I saw SQL Server on Kubernetes for the first time. I saw firsthand the benefits of administration, system management, and also the development speed provided by running SQL Server in containers on Kubernetes. At that time, I chose to invest my own learnings in running SQL Server on containers and also Kubernetes. I strongly feel that this is the future when it comes to deploying SQL Server.

In this chapter so far, development languages and frameworks, tooling, platform management, and also platform options, such as IaaS VMs and containers, have all been introduced. This can get to be a bit overwhelming, so let's take some time to discuss how to choose a platform for deploying SQL Server in Azure.

So many choices: which platform should you choose?

So far in this chapter, we have outlined several deployment scenarios for SQL Server on Linux in Azure. In this section, you will understand your platform choices and some of the decision points to help you decide which is best for your scenario. As a developer, it's important that you consider the platform your application will run on.

Which base operating system?

When using IaaS VMs, you have the choice of selecting the base operating system you want to deploy. While deciding the base system, there are two key elements: the technical requirements and the skills of the organization. The technical requirements of the application are something that can surface pretty easily during the analysis phase of building a system architecture. In this section, I'm going to focus on operations and costs—two things that are often forgotten when developing system architectures.

Operations and maintenance

When working with VMs, key operations and maintenance are going to still be under your control. So, things such as software installation, patching, on-going system maintenance, and performance-tuning are still your responsibility. This is fine if that's the requirement for your deployments and your organization. These elements are going to be key to choosing the base operating system. If your organization is one that is familiar with administering Red Hat-based solutions, you can leverage your existing skills and tooling to manage your Azure IaaS VM-based platform. If your organization is one that is skilled with Ubuntu and the universe around this distribution, then you have this as a choice as well. The key concept is that the platform is your choice to make.

You can learn more about the most recent performance best practices for Linux from this Microsoft documentation[39].

Operating system costs and support options

In Azure, the Marketplace images that are available with SQL Server on Linux pre-installed are based on Red Hat Enterprise Linux, SUSE Linux Enterprise Server, and Ubuntu. Red Hat and SUSE are licensed software products that have "pay as you go" and "bring your own software" offerings in Azure. Ubuntu's offering in Azure comes at no additional licensing costs for the operating system, but this free edition does not include support—support is available for purchase from Ubuntu's publisher, Canonical. Production support or enterprise agreements can be a critical factor in your deployment scenario and can impact your choice of base operating system based on the needs of your organization.

How should you choose between containers and VMs?

You may have heard that containers are the new VMs and we described earlier in this chapter the benefits of using containers and also how you can combine IaaS VMs and containers in Azure. Furthermore, modern application and infrastructure architectures are trending toward container-based deployments in container orchestrators such as Kubernetes and OpenShift. These deployment scenarios enable private cloud and serverless architectures.

The focus of this chapter is deploying SQL Server on Linux in Azure IaaS VMs, and a primary motivation for running SQL Server in containers on IaaS VMs is to support multi-instance scenarios. SQL Server on Linux does not have the concept of a named instance as in the case of SQL Server on Windows. So, to run multiple instances of SQL Server on a single base operating system or a single VM, you need to use containers.

From a deployment standpoint, you do have to take into consideration the fact that you are sharing the resources of the underlying VM. So, in terms of CPU, memory, and disk, you will need to ensure that your container-based SQL Server workloads have sufficient resources and that these are shared and balanced accordingly among the containers running your SQL Server workload.

Finally, when running SQL Server in containers, for ideal performance, you should run the same operating system inside the container as you do on the base operating system. This will ensure that there are no virtualization technologies in play when running your SQL Server workloads, thereby yielding maximum performance.

Why should you do this in Azure?

The majority of this chapter has been about running SQL Server in Azure and in VMs and containers. Let's have a look at some of the common Azure services that can be combined with Azure-based SQL Server solutions and building and deploying enterprise applications in Azure:

- **Azure Load Balancer**: Used to distribute loads into multiple application servers for scale-out performance and redundancy.

- **Azure Traffic Manager**: DNS-based load balancing between Azure regions. Used for scaling workloads beyond a single region and also for high availability and disaster recovery between regions.

- **Azure Site Recovery**: Provides business continuity and disaster recovery of applications at the VM level between Azure regions with coordinated, dependency-aware failover.

- **Azure Recovery Services Vaults**: Provides VM-level snapshot-based backups and replication of those backup vaults between Azure regions.

- **Azure Storage Accounts**: Using Azure Files, which is a feature of Azure Storage accounts, you can expose a CIFS mount in a Linux VM. This can be used as a SQL Server backup target. SQL Server backups are stored in Azure Files and can be protected with snapshot backups and replication between Azure regions. Azure Files is one of my "go to" solutions for customers who need to manage SQL Server backups in the cloud, leveraging the benefits of snapshots and inter-region replication to provide a high level of data protection.

- **Azure Security Center**: Azure Security Center is a unified infrastructure security management system that strengthens the security posture of your datacenters and provides advanced threat protection across your hybrid workloads in the cloud. In an IaaS environment, you need access to the tools that Azure Security Center provides to harden your network, secure your services, and make sure that you're on top of your security posture.

Summary

In this chapter, we introduced the open-source development ecosystem around SQL Server on Linux and how SQL Server can help add business value to applications. We introduced the tools available for you to build and manage your data estate in Azure, including the Azure CLI and Azure Data Studio. We then looked at the platforms available to run SQL Server on Linux in Azure and some of the decision points to focus on when selecting your application's platform. We then wrapped things up with a discussion of some of the most common Azure services that can be used to build a data platform in Azure. With that, let's move on to the next chapter, which will focus on the performance of SQL Server, best practices, and how you can optimize workloads running in SQL Server on Linux in Azure.

Chapter links

1. https://bit.ly/2TG5aDZ

2. https://bit.ly/3ca0IUF

3. https://bit.ly/3enigOz

4. https://bit.ly/2X3gCMb

5. https://bit.ly/2XyWoJ0

6. https://bit.ly/2ZBIdWl

7. https://bit.ly/3glUZhC

8. https://bit.ly/3c3vWfY

9. https://bit.ly/2X3kTz4

10. https://bit.ly/2ztxBhu

11. https://bit.ly/2ZIvkKl

12. https://bit.ly/3gmYHHI

13. https://bit.ly/3c4ExPA

14. https://bit.ly/3c4Ezaa

15. https://bit.ly/3ekRhTO

16. https://bit.ly/2X2MIYs

17. https://bit.ly/3enW1HZ

18. https://bit.ly/2LZshVP

19. https://bit.ly/3gqB4xZ

20. https://bit.ly/2ZBLsgt

21. https://bit.ly/3eq6mU9

22. https://bit.ly/2X1j85k

23. https://bit.ly/3en9Y9s

24. https://bit.ly/2ZEYEBc

25. https://bit.ly/36vRgcM

26. https://bit.ly/2ZDQJnB

27. https://bit.ly/3ew0jgT

28. https://bit.ly/2Ab3fAk

29. https://bit.ly/2ZDGv70

30. https://bit.ly/2A40ReN

31. https://bit.ly/2XtauvD

32. https://bit.ly/2LYfLG3

33. https://bit.ly/2zvhy2P

34. https://bit.ly/2X1kqgG

35. https://bit.ly/2TEPKzO

36. https://bit.ly/2LY5z0j

37. https://bit.ly/2X3mdSH

38. https://bit.ly/2XxFWsq

39. https://bit.ly/2Xtc7tf

5

Performance

By Tim Radney

The previous chapters have given you an understanding of SQL Server, the overall benefits of the various capabilities of SQL Server on Azure VMs, and how to get started with SQL Server on Azure VMs.

In this chapter, we will discuss SQL Server performance best practices and how you can achieve the best performance for your SQL Server workload. This chapter will be divided into three main parts:

- Performance best practices

- How to optimize SQL Server on Azure VMs

- Azure BlobCache

Let's begin by taking a look at the best practices to follow in order to get the most from your SQL Server.

Performance best practices

Performance tuning can be a broad and complex topic; however, migrating to Azure VMs can help simplify the process. There are so many factors that can impact performance. If you have already deployed an Azure VM, the Azure portal lets you use Azure Monitor for VMs. It provides insights into the health and performance of your Windows or Linux VMs by monitoring their processes and dependencies on other resources. Not only can it monitor other Azure VMs; it can monitor VMs on other cloud providers and on-premises to give you a holistic view. You get pre-defined performance charts that show the trending and the dependency map, all built into the Azure portal. In addition to Azure Monitor for VMs, all VMs have basic monitoring enabled in the Azure portal. This basic monitoring shows the average CPU, the total network, the total disk bytes, and the average disk operations per second. Data can be shown for the past 1, 6, or 12 hours; the past 1, 7, or 30 days; and can be found in the Overview tab of the VM.

When planning a new deployment, upgrade, or migration for the SQL Server workload, one of the first considerations is the size of the server that will handle the workload. Regardless of the server environment, whether it's physical or virtual, on-premises or on the cloud, the amount of CPU, memory, and storage will always be factors that influence the deployment. Azure VMs help here by having specific VM types for different workload needs:

- **General purpose**: Balanced CPU-to-memory ratio. Good for small- to mid-size databases.

- **Compute optimized**: High CPU-to-memory ratio. Good for ETL servers.

- **Memory optimized**: High memory-to-CPU ratio. Great for relational database servers, medium to large caches, and in-memory analytics.

- **Memory optimized constrained vCPUs**: The vCPU count can be constrained to one half or one quarter of the original VM size. This reduces the SQL Server license cost and allows you to have a low CPU-to-memory ratio.

- **Storage optimized**: High disk throughput and I/O. Ideal for big data, SQL, NoSQL, data warehousing, and large transactional databases.

- **High performance compute**: Designed for leadership-class performance, MPI scalability, and cost efficiency for a variety of real-world HPC workloads.

Whatever your workload, there is a VM type for you. There are other considerations for performance as well, such as the configuration of SQL Server and the overall maintenance of the system.

In the following sections, we will begin by discussing these factors, starting with VM storage.

Virtual Machine Storage

Azure VMs are an excellent choice for running SQL Server workloads. *Chapter 2, Getting Started with SQL Server in Azure Virtual Machines*, introduced you to getting started with Azure VMs and *Chapter 4, SQL Server on Linux in Azure Virtual Machines*, covered the various Linux and Windows distributions and their advantages. The disk types were explained, and you should understand that production SQL Server workloads should be running on premium SSDs or ultra disks. Most SQL Server workloads are I/O intensive, meaning production workloads require faster storage. Premium SSDs are designed for production workloads and workloads that are sensitive to performance. For very intense I/O workloads, ultra disks are a better choice as they deliver higher throughput, higher IOPS, and lower latency. Consider ultra disks when you have a very transaction-heavy workload. When considering CPU, memory, and storage, storage is typically the slowest component, which means that faster storage is better.

When it comes to selecting a VM, the size and the type matter for numerous reasons. The number of disks a VM can have is tied to the size of the VM in addition to overall IOPS and throughput. Organizations generally struggle with selecting the right size of the VM to handle their I/O needs. It is entirely possible to configure a VM with a storage solution that can provide a high number of IOPS and throughput that the VM throttles and hits the ceiling of what the VM supports. In this type of situation, the VM will have to be scaled up to a larger size that supports a higher number of IOPS and throughput to have better storage performance. If you are configuring the VM through the Azure portal and select a storage configuration that exceeds the throughput limit of the VM, a message is displayed to warn you about this limitation.

When it comes to sizing a VM for an existing workload that is being migrated to Azure, it is common to try to match the existing memory, CPU, and storage capacity. What is needed in order to properly size an Azure VM is a baseline. A baseline is simply a point of reference that is measured. When working with SQL Server, baselines are important so that you can capture when there is a change in behavior or performance. Common things to capture are CPU and memory utilization, disk latencies, and batch requests per second, among other things. Having baselines is important; you need to know what normal behavior looks like so that you can tell when things are different. A baseline for existing I/O and throughput usage is critical in order to know what the maximum throughput is that you'll need the Azure VM to handle.

You can use Performance Monitor[1] to capture physical disk metrics; however, the `sys.dm_os_virtual_file_stats` **dynamic management view** (**DMV**) provides the key information needed to calculate overall throughput by returning I/O statistics for data and log files. The `num_of_bytes_read` and `num_of_bytes_written` columns return the number of bytes read and written to each file. If you capture the results of this DMV over a short period of time, you can calculate the MB/s for reads and writes to know whether the size of the VM with the storage you provided can support this workload. This is a crucial step that many organizations miss when migrating SQL Server workloads to the cloud and then experience slower overall performance.

As an alternative sizing method, you can use the Azure Migrate: Server Assessment tool. Azure Migrate collects real-time sample points for a month and analyzes the data to identify the ideal VM size based on those metrics.

It is common for on-premises VMs with lower logical core counts to have very high throughput capabilities to the local storage area network (SAN). In a public cloud scenario, constraints must be in place to prevent one VM from consuming all the storage throughput on that host. It makes sense that the available throughput and I/O to the host is allocated based on the size of the VM. The larger the VM, the more resources it is assigned. This must be a consideration for performance when planning a migration to Azure.

To handle the I/O requirements, you should only consider Azure premium disk or ultra disk for your production SQL Server environment.

Memory

If you ask a **Database Administrator** (**DBA**) how much memory a SQL Server needs, a common response is "more" or "all of it". The reason why it is good to allocate SQL Server a lot of memory is that the buffer pool will store recently read data in memory. If that data is accessed again while it is in memory, SQL Server doesn't have to go back to the disk to retrieve it, which is a slow operation. These are called logical reads when it pulls from memory.

For SQL Server workloads that need high memory-to-CPU ratios, you should choose the memory-optimized type of VMs. In many cases, you may need even more memory than offered by the standard memory-optimized type of VM. For those situations, the memory-optimized constrained vCPU VM offers the best price for performance. It allows you to select the much larger vCPU and memory VM and only use a quarter or half of the available vCPU. This gives you higher memory, storage, and I/O bandwidth without the higher cost of SQL Server licenses.

Having more memory also opens up the ability to take advantage of memory-optimized tables known as In-Memory OLTP. In-Memory OLTP is designed to optimize the performance of transaction processing, ingesting data, loading data, and various transient data processes. In-Memory OLTP is just what it sounds like. It allows you to create memory-optimized tables and store data into memory. It gets its performance gains by removing lock and latch contention between transactions, making transaction execution much more efficient. For very volatile or large-volume transaction processing, In-Memory OLTP can drastically improve performance. Learn more about In-Memory OLTP here[2].

Another SQL Server feature that can be utilized is columnstore indexes. Columnstore indexes provide a different way of storing data that can provide performance improvements for certain types of queries. A columnstore is data that is physically stored in columns and then logically organized into rows and columns. The columnstore index slices the table into rowgroups and then compresses them. With the high level of data compression, typically by 10 times, you can drastically reduce the storage cost and improve I/O bottlenecks in your environment. Columnstore indexes can be beneficial in both OLTP and data warehouse environments. Columnstore objects are cached in the columnstore object pool instead of the SQL Server buffer pool. There is a 32 GB limit for the columnstore object pool in the Standard Edition, whereas there is no limit for the Enterprise Edition.

If a system starts suffering from memory pressure, organizations can easily scale an Azure VM up or down to another size that has more or less memory as needed. This ability to scale up and down as workloads change is one of the biggest benefits of Azure VMs and can save organizations a ton of money. To have that benefit on-premises, you have to own the hardware that you may only need a few times per year for scale, whereas in Azure, you can scale as needed. I've worked with numerous clients over the years who have a seasonal business. Open enrollment for healthcare, registration for universities, online retailers for Black Friday, and so on require extra compute during their busy seasion. They can scale up and just pay for the additional compute during that season and then scale back down. For example, a customer may have D16s_v3/D16as_v4 with 16 vCPUs and 64 GiB of RAM. Their busy season is approaching and they know their workload will increase by 40-50%. They could simply scale to D32s_v3,/D32as_v4, which would double their vCPUs from 16 to 32, their memory from 64 GiB to 128 GiB, and also their temporary storage from 128 to 256 GiB. Since they are staying within the same VM series, all that would be needed is a restart. If the resize is to a different type of series, or the hardware cluster hosting the VM does not support the new VM, the VM will be moved to a new host, which can take more time.

For existing SQL Server instances that are being migrated to Azure, we strongly recommend analyzing the current memory utilization before assuming that you need the same memory post-migration. If data is staying in the buffer pool for excessive amounts of time, you could consider a VM with less memory. For example, I had a customer who was on a 4 vCPU on-premises VM that was allocated 128 GB of memory. I was measuring page life expectancy (how long data pages stay in memory) in days, not minutes or hours. They were not using In-Memory OLTP and it was a dedicated machine just for SQL Server. We were able to migrate to a VM with less memory, saving the customer a lot of money.

At the same time, if you are experiencing memory pressure, tuning and optimizing should be considered before migrating to decrease the memory pressure. Properly sizing the Azure VM is necessary for better SQL Server performance, but also to ensure that you are not overspending on resources that you don't need.

CPU

Selecting the number of processors can have a direct impact on SQL Server performance if you underestimate the amount of CPU needed. If you overestimate the number of processors, you won't see a performance improvement; however, it will affect your SQL Server license cost. Workloads change over time, and at some point more CPUs may be needed to handle that workload. Microsoft has published a list[3] of the Azure compute units (ACUs) for the different VM families. This gives you a quick reference to compare CPU compute performance across Azure SKUs.

By running SQL Server on Azure VMs, you can easily scale to a server with more vCores, which typically also includes more memory and more I/O throughput. Don't forget about those constrained vCPU options if you have a small vCore requirement but a need for larger memory. However, if you have a high compute need, compute-optimized VMs offer a higher core count. For low CPU latency and fast clock speeds, consider the Eav4 VM series, which features the AMD EPYC™ processor.

Properly sizing the Azure VM has a direct correlation to having the foundation for a SQL server that performs well. The number of CPUs needed, the amount of memory required, the overall storage capacity, and the storage I/O are all factors that control the type and size of Azure VM you need. Each factor, by itself, can force you into a certain size VM or VM type. Gathering these requirements early on is a big factor in having a SQL server with good performance, and a successful deployment.

SQL Server configuration

TempDB is a system database that is utilized by many processes for storing work tables, temporary tables, spills, row versions, and much more. TempDB is a unique database due to its characteristics. For example, there is only one TempDB database for the entire SQL Server instance, it's recreated when the SQL Server services are restarted, and technically, TempDB cannot be backed up. However, TempDB is a mission-critical database for the SQL Server instance and can become a place of contention for SQL Server. For that reason, TempDB needs to be properly sized for the instance and requires more than a single data file. In most gallery images, there is a single data file and a log file for TempDB. With TempDB being utilized by the entire instance, contention can develop on the pages related to Page Free Space (PFS), Global Allocation Map (GAM), and Shared Global Allocation Map (SGAM). These are pages 1, 2, and 3 in the data file. A new PFS page is created every 64 MB, and a new GAM and SGAM page is created every 4 GB.

To alleviate the potential for contention on these pages, you need more data files. The generally accepted rule is one equal size data file per core up to 8-cores. If you have more than 8-cores, start with eight data files of an equal size. If contention is still an issue with eight files, create more equally sized data files in increments of four. You should also set the initial size of the data and log files to be the size that TempDB grows to after an initial workload. For example, if you have a 4 vCore server and TempDB grows to 8 GB in size after a normal workload, you would need to create four TempDB data files with an initial size of 2 GB each and make sure each is set to auto grow by the same fixed size.

The default auto growth size for TempDB on SQL Server 2017 and SQL Server 2019 is 64 MB. This can be changed to a higher value if needed. Beginning with SQL Server 2016, when TempDB data grows, each TempDB data file grows at the same time. Prior to SQL Server 2016, this was controlled using trace flag 1117. Also beginning with SQL Server 2016, when an extent is created, all eight pages are created at the same time. Prior to SQL Server 2016, this was accomplished using trace flag 1118.

A common practice is to take advantage of the local SSD that is part of every Azure VM for storing TempDB. VMs have varying sizes for this local disk, so you'll have to make sure your local SSD is of proper size before utilizing it. Memory- and storage-optimized VMs offer a higher capacity local SSD storage than the general-purpose ones. Depending on the TempDB utilization, isolating TempDB to its own premium or ultra SSD may be needed. In some extreme cases, isolating different TempDB data files to a separate disk may also be needed to further distribute I/O. See *Appendix* A for information on SQL Server configuration with OLTP workloads.

Dynamic management views (DMVs) and Query Store

SQL Server on an Azure VM is still SQL Server, regardless of Linux or Windows. Standard query tuning is still required to make SQL Server run as well as possible. This means monitoring for queries that are consuming the most resources. Common approaches are looking for long-running queries, queries that are executed the most, and those consuming the most CPU and disk I/O.

SQL Server 2005 introduced DMVs as well as dynamic management functions (DMFs). Every version of SQL Server since has introduced new and improved DMVs and DMFs to help manage and support SQL Server. Some common DMVs for performance tuning include:

- `sys.dm_os_performance_counters` – returns SQL Server performance counters
- `sys.dm_db_index_usage_stats` – provides detailed usage of indexes to include user seeks, scans, lookups, and more
- `sys.dm_exec_cached_plans` – provides all cached query plans currently available
- `sys.dm_exec_query_plan` – provides the show plan in XML of the query plan
- `sys.dm_exec_query_stats` – returns stats for cached query plans
- `sys.dm_exec_sql_text` – returns the text of the query
- `sys.dm_io_virtual_file_stats` – provides I/O statistics that can provide latency and throughput usage
- `sys.dm_os_wait_stats` – returns information about what SQL Server is waiting on
- `sys.dm_exec_sessions` – returns one row per session
- `sys.dm_exec_connections` – returns the details of each connection
- `sys.dm_tran_active_transactions` – provides the transaction state for the instance

There are numerous categories of DMVs, and DMVs can be joined with other tables and DMVs to create robust queries. Most DBAs have a collection of DMV scripts to use to collect system information on their SQL Server instances, as well as for troubleshooting performance issues when they arise.

SQL Server Query Store provides insights into the query plan choice and performance. It automatically captures the history of queries, plans, and runtime statistics and retains them for review. You can read more about Query Store in *Chapter 3, Hero capabilities of SQL Server on Azure VMs* .

How to optimize SQL Server on Linux

There are several Linux-specific changes that need to be made for SQL Server deployments on Linux. Some changes may be specific to certain Linux distributions.

An administrator should disable the last accessed date/time (atime) on any filesystem that is used to store SQL Server data and/or log files. The last accessed date/time mount option causes a write operation to happen after each read access. This would generate a massive amount of extra I/O. The mount option should be changed to noatime to help reduce disk I/O.

In order to deal with large amounts of memory, Linux uses **Transparent Huge Pages (THP)**. THP automates managing, creating, and working with huge pages. Manually trying to manage huge pages would be very difficult, so, for SQL Server on Linux, leave THP enabled.

Memory should be managed so that SQL Server does not starve the underlying operating system. At the same time, SQL Server should be configured to use as much memory as possible without causing an issue with the operating system. SQL Server, by default, will only use 80% of the physical memory. If the remaining 20% is too significant and wasteful, you can manually configure the value. This is controlled by the `memory.memorylimitmb` setting. To change the value, use the `mssql-conf` script and set the value for your server to the `memory.memorylimitmb` value. You should also have a configured swapfile in place to avoid any memory issues.

Azure BlobCache

When configuring the VM or adding storage, you can set the caching for the disk. BlobCache provides a multi-tier capability by using the VM's memory and the local SSD for caching. This is only available for Premium disk and is the default for Azure Marketplace images. If you've self-installed SQL Server, then you'll have to manually set up Azure BlobCache.

There are three options for disk caching: ReadWrite, None, and ReadOnly.

For SQL Server workloads, ReadWrite should not be used. ReadWrite can lead to data consistency issues with SQL Server.

None should be used for SQL Server log files. SQL Server log files write data sequentially to disk. There would be no benefit to using ReadOnly caching in that scenario.

SQL Server data files should take advantage of the ReadOnly caching option. By using ReadOnly, reads are pulled from the cache, which are stored within the VM memory and the local SSD. Memory is fast, and the local SSD provides better performance than the remote SSD storage. What is even more crucial is that the reads from the cache are not counted towards the disk IOPS and throughput limits. This allows you to achieve higher overall throughput from the VM. ReadOnly caching provides lower read latency, higher throughput, and overall higher read IOPS compared to premium remote SSD.

If you are taking advantage of ReadOnly caching for SQL Server, reserve the local temporary SSD for caching, and do not place TempDB on this disk. Leave the I/O for caching and place TempDB on its own premium or ultra SSD.

Summary

For great performance with SQL Server on an Azure VM, you need to first size the VM for your workload, paying attention to CPU, memory, and storage I/O capabilities. You can easily select the right configuration for your workload with the vast array of Azure VMs available. Many have been configured to provide the best price for performance for SQL Server workloads. By leveraging the highest generation VM sizes with Azure BlobCache and ultra disks, you can run just about any size SQL Server workload on an Azure VM. To know what size VM you need, you'll need a baseline, and to have SQL Server run as smoothly as possible, you'll want to ensure that you've configured it properly for your workload. In addition to SQL Server configuration settings, there are key changes to make in Linux environments, especially setting noatime to reduce wasteful I/O.

Chapter links

1. https://bit.ly/2X3b8Ba
2. https://bit.ly/2X2nwRV
3. https://bit.ly/36yDV3r

6

Moving workloads to SQL Server on Azure Virtual Machines

By John Martin

In the previous chapters, we covered the core elements of running Microsoft SQL Server in Azure VMs. Now that we understand how to provision and configure our VMs, we will move our on-premises workloads to the cloud.

Moving on-premises workloads directly to Azure can be daunting. However, by making use of Azure **Virtual Machines (VMs)**, we can fully realize the potential of cloud-based agility and versatility without needing to re-engineer our applications. This approach also helps us when we face challenges such as refreshing aging hardware or upgrading to supported SQL Server versions. What historically could have taken months to plan and execute with high up-front spending, can now be handled without the initial expense in a much shorter timeline.

We have two main options for handling the migration process to Azure VMs. We can tune the workload ahead of moving it to the cloud, which can prove to be cost-efficient but will take longer to realize the value. Or, we can move the workload and then tune it. The latter approach allows us to make use of cloud capabilities and new SQL Server features such as Query Store, which accelerates our workload tuning.

Achieving benefits of the cloud such as increased agility and scale is possible by lifting and shifting workloads to Azure VMs or cloud modernization to an Azure SQL Managed Instance. The exact target will be dependent upon the workload and its dependencies as to which is the best fit initially. In addition to the benefits mentioned already, there are several additional benefits regarding high availability and disaster recovery, which can be leveraged simply by using the Azure cloud.

In this chapter, we will look at some of the tools and techniques available for lifting and shifting our workloads in the most time-efficient and cost-effective manner. This chapter will be divided into three main parts:

- Migration tools and best workload migration practices
- Application considerations
- Moving to the Power BI service from Power BI Report Server

The most important element of any migration is to effectively identify and locate data/workload that needs to be migrated. In the past, this had to be done manually, but now there is a great array of tools that we can make use of to achieve this.

Migration tools and best practices

There are many ways to get your SQL Server workload from on-premises to Azure VMs, but which one should you choose?

In this section, we will look at the best practices to follow as well as tools available to us in order to successfully migrate on-premises workloads to Azure VMs. Let's begin by looking at best practices.

Best practices

When migrating workloads to the cloud there are a few key factors that need to be considered in order to achieve success. These are:

- Understanding the scope of what is impacted by the migration.
- Identifying the current performance levels.
- Timelines for migration activities.

By understanding these key requirements, we are then able to plan and implement the migration of workloads.

Not having a full picture of what needs to be included in a workload migration is one of the most common hurdles that we need to overcome. This is the foundation to successfully accomplish all of the migration activities. If we do not get this right, it is akin to building a skyscraper on sand.

Defining migration scope

There are three main elements to this scoping exercise:

1. Identifying systems that are to be migrated and the ones that are not to be migrated.
2. Technical analysis of the components within the system.
3. Speaking with the consumers of the system to understand their needs.

First and foremost, we need to understand and identify the servers and databases that we will be moving. Are the databases currently co-hosted with other databases on the same server or are they hosted on a dedicated server? Understanding this will help us get the information that we need to prepare the target server.

Once we understand which systems are in scope, we need to analyze the source system(s). Identifying both instance- and database-scoped configurations and objects is key. Everything ranging from linked servers, credentials, and logins, to SQL Agent jobs will need to be assessed to identify whether they need to be migrated or replaced on the new platform.

Finally, by speaking with the consumers at this time, we can understand the perceptions around the current performance and availability requirements. The availability requirements help us define the migration process for the database(s) that need to be moved.

Identifying current performance

An age-old problem with migrating workloads from one system to another is the dreaded "It's not performing like it used to" statement from the consumers. The best way to tackle this is to ensure that you have solid benchmarks and performance details for comparing workloads post-migration.

> **Note**
>
> When looking to understand SQL Server performance, you should really base your monitoring and analysis on the Waits and Queues methodology as seen in this Microsoft whitepaper[1].

By using the **Waits and Queues** approach we can rapidly identify the key performance counters we need to track. In the queues category, we can capture counters that fall into three key pillars. Underpinning these is the foundation of what we can think of as "general" counters, which we can use to track workload throughput.

Figure 6.1 highlights some of the key performance metrics (queues) that we need to consider when understanding our SQL Server workloads.

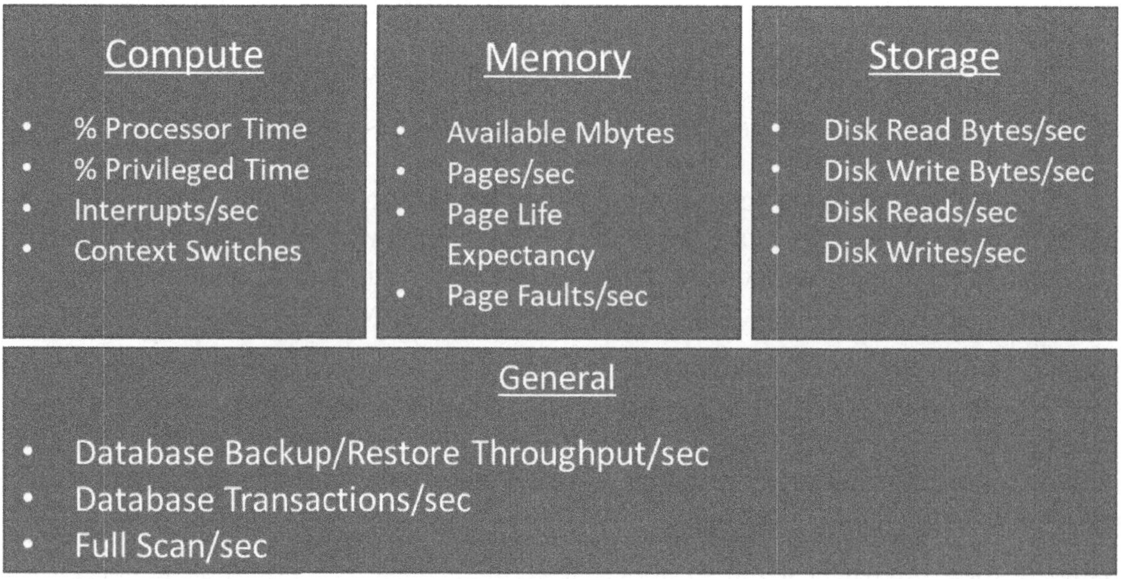

Compute	Memory	Storage
• % Processor Time • % Privileged Time • Interrupts/sec • Context Switches	• Available Mbytes • Pages/sec • Page Life Expectancy • Page Faults/sec	• Disk Read Bytes/sec • Disk Write Bytes/sec • Disk Reads/sec • Disk Writes/sec

General
• Database Backup/Restore Throughput/sec • Database Transactions/sec • Full Scan/sec

Figure 6.1: Key performance metrics

> **Note**
>
> It should be noted that **Page Life Expectancy** (**PLE**) on its own is not a key performance indicator for your workload. It is important to understand the size of the buffer pool and how much of your I/O sub-system capability is being consumed to maintain the PLE number to dictate what is appropriate and what is not.

When these key performance metrics are combined with server-level wait statistics collection, it will give us a good idea of whether there are any pain points. Additionally, this capture will act as a performance baseline that we will use to validate the new system's performance.

Migration timeline

Defining a timeline for migration activities is vital but it needs to be realistic. You may come across a number of situations where the timelines have been set with unrealistic goals. This results in us needing to cut corners and delivering an inferior customer experience. Key factors that we should consider when setting migration timelines are:

- System complexity.

- Platform and application validation.

- Volume of data to be migrated.

- Acceptable downtime for the business.

These four elements will help us create an estimated timeline for a successful migration process. Within these phases, it is important to identify where we can run parallel tasks as well as to make use of automation for repeatable tasks.

Automation is key when performing migration activities. Everything ranging from the initial analysis to platform validation, and ultimately the migration itself, allows us to perform consistent and repeatable database migrations.

Migration and analysis tools

Microsoft provides several great tools for helping us move on-premises workloads to the cloud. They range from ones that help us identify blockers through to those that move our databases. The tools that we are going to look at here are:

- Microsoft Assessment and Planning (MAP) Toolkit[2]

- Data Migration Assistant (DMA)[3]

- Database Migration Service (DMS)[4]

In addition to these, there are also some key community-based open-source tools that we can use to speed up the process and automate the migration of workloads, including:

- dbatools[5]

- WorkloadTools[6]

The tools listed above fall into two categories: analysis and migration. Some of them can perform or facilitate both such as the **Data Migration Assistant (DMA)** or dbatools, but that is not the primary purpose.

One of the most common challenges we will face with any migration is not understanding the requirements of our workloads. There is a common misconception that because we are moving from on-premises servers/VMs to cloud-hosted VMs, we don't need to do our homework. Although the operating system and SQL Server that we have been running on-premises are the same, the underlying platform and the overall operating model is radically different.

MAP Toolkit

Our first port of call is to start with gathering key information about our existing environment. The quickest and easiest way to perform this is with the MAP Toolkit from Microsoft. It is a free tool that can discover and analyze SQL Server systems. This has the advantage of letting us scan one or more servers and discover a lot of detail around the SQL Server elements that are installed on our source systems.

Figure 6.2 shows how the MAP Toolkit can help us identify the installed components on the source servers. This will help us understand whether we can simply lift and shift to the cloud or need to augment with additional services such as Azure Data Factory or Power BI.

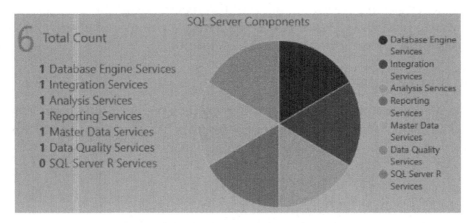

Figure 6.2: MAP database analysis

Using MAP database analysis, we can capture details for the configuration of the server and databases by generating reports from the MAP Toolkit. This information then allows us to define the configuration of the target systems.

Data Migration Assistant (DMA)

The DMA takes our analysis to a deeper level and starts assessing whether we have any blockers. It also allows us to view any recommendations that we might want to address ahead of moving our database.

DMA is designed to perform an analysis of many different combinations. We are going to start with SQL Server as the source and target SQL Server on Azure VMs. The process for configuring DMA to assess databases and servers can be seen in *Figure 6.3*.

> **Note**
>
> While having a high degree of parity, SQL Server 2019 on Linux does not have all the same features as the Windows version. If we are targeting SQL Server on Linux, the DMA helps us spot any potential incompatibilities without us needing to review the documentation manually.

There is a logical flow to using DMA for analysis, which will give us a clear and concise report about any potential blockers. Before we run the analysis, we need to understand what we want to know and where we are going from and to. We can then get into the specifics of which tests we want to run as well as the specific versions of SQL Server. One of the useful features in the DMA is it's ability to load in workload trace files for the database(s) we are analyzing. And we would encourage you to do this to get a better picture of what is going on. Once all of this is complete, you can run the analysis and then review the report.

> **Note**
>
> It is possible to automate the use of DMA with command-line arguments. More details can be found here[7].

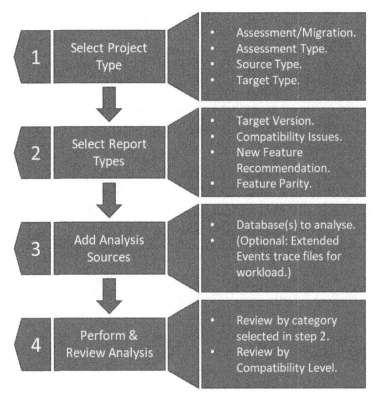

Figure 6.3: DMA analysis steps for the SQL Server VM target

The subsequent DMA report highlights legacy T-SQL syntax and other incompatibilities, deprecated and discontinued features, and recommendations for feature usage in the target version. For example, it can identify potential objects where dynamic data masking or Always Encrypted column-level encryption features could be beneficial based on the object names.

While DMA has a primary role in analysis, it can also perform basic database migrations. However, for a richer and more robust migration experience, we would suggest using Azure Database Migration Service.

WorkloadTools

The final piece of the puzzle to successfully plan moving workloads to Azure is the workload analysis.

Here, we would recommend you look at an open-source tool called WorkloadTools. This tool can capture, analyze, and replay workloads. This capability means that we can ensure that a migrated workload will behave in the expected way once it has been moved to Azure.

By using WorkloadTools, we can mirror production workloads in an Azure VM. This allows us to demonstrate that the new system will meet the performance requirements, while at the same time allaying any fears that you might have about the new platform.

Figure 6.4 illustrates the high-level architecture to configure the WorkloadTools system to collect, replay, and compare workload details. The first step is to capture the workload metrics from the source system to be migrated. This is replayed to a test copy of the database on the target infrastructure configuration. The second step is to collect the performance metrics from the replayed workload. Both activities store their performance data in a central database in different schemas. This facilitates step 3, which is where we review and compare the metrics from the two captures to ensure that we have a comparable level of performance. If not, we can then reconfigure the target system and rerun our tests and captures.

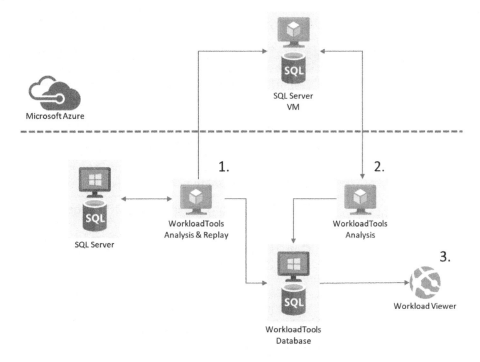

Figure 6.4: WorkloadTools architecture

Once we have finished our analysis, we can move the database and its workload. We can also then reuse these tools to perform continued troubleshooting and workload analysis through the lifetime of the system. By storing the data each time, we can track improvements or regressions over time.

SQL Server is a highly versatile and adaptable database engine with a mature tooling ecosystem. This means that we have a number of different ways to move databases to the cloud.

One of the most common methods for migrating databases between SQL Server systems is to perform a backup and restore. However, depending on the size of the database, this can range from a very short time to many hours for multi-terabyte databases. We can look at options such as log shipping, transactional replication, or availability groups to perform the migration with less downtime. But these will depend on a number of external factors that could complicate matters. We will look at how to migrate databases by manually taking a backup and restoring it to the target. We will also see how to automate this with log shipping and then touch on **Database Migration Service (DMS)** and DMA to migrate databases. But first, we need to get the prerequisites moved at the server level.

Key analysis points

Once we have completed our analysis phase, we will be in a position to move forward to the migration planning. However, in order to plan effectively, we need to have collated our analysis data so that it is able to inform the following key decision points:

Analysis Point	Details
Migration Blocker	Deprecated, discontinued, or unsupported features that prevent migration to Azure Virtual Machines.
Business Requirements	Disaster recovery (RPO/RTO), migration downtime windows, cost, and licensing.
Resource Usage	CPU, memory, I/O, storage volume, network bandwidth,and latency.
Workload Requirements	Parallel query processing, TempDB usage, OLTP versus DSS, and collation.
Security Requirements	Geo-locality for data, encryption requirements, user access, and RBAC.
Application and User Experience	Application response times, transaction throughput, and query response times.
Maintenance Operations	Maintenance windows, job runtimes, and resource usage profiles.

Figure 6.5: Key decision points for migration planning

Once we have these key data points, we can build out an information pack, which we will be able to refer back to during our migration.

When reviewing the output from this analysis phase, we need to do so with a cloud mindset. This means that whereas historically we would buy and provision servers with lots of headroom to grow, instead now, we want to run our servers hot. The ability to easily scale up our servers without needing to first put physical resources in place is one of the great capabilities of cloud-based VMs. For example, we should review our CPU resource usage; if it is 40% with spikes to 60% on-premises then when we move to the cloud, we can look to pick a machine with less compute that results in us running at 60% with spikes to 80%. This will help us get a better ROI on the cloud infrastructure we deploy.

Another area that many people neglect to review properly is the maintenance window activity. Database consistency checking, index and statistics maintenance, and backups are all still applicable in the cloud. Ensuring that you have the appropriate resources in place to perform these activities in the time windows available is important.

dbatools—migrating instance-level objects

While moving the database takes much of our focus, a key prerequisite is that instance-scoped objects are migrated first. If we do not move our logins, credentials, linked servers, Agent Jobs, etc., then we will not have a successful migration as there are database dependencies on many of these. There are many ways to perform these tasks—from custom T-SQL to SSIS and even bespoke SMO application code. However, we would recommend using the dbatools community PowerShell module for these activities.

Within dbatools there are several commands that are dedicated to the migration of SQL Server objects. What this means is that we can write PowerShell scripts to perform these migration activities with the added benefit of then leveraging automation technologies to execute them.

The key migration commands that we should look at understanding and using are:

`Copy-DbaAgentAlert`

`Copy-DbaAgentJobCategory`

`Copy-DbaAgentJob`

`Copy-DbaAgentOperator`

`Copy-DbaAgentProxy`

`Copy-DbaCredential`

`Copy-DbaLogin`

`Copy-DbaCustomError`

`Copy-DbaDbMail`

`Copy-DbaLinkedServer`

`Copy-DbaXESession`

`Copy-DbaSpConfigure`

`Copy-DbaAgentSchedule`

There is also the **Start-DbaMigration** command within the PowerShell module. This will perform many of the key activities to move instance-scoped items, including all the items listed above, as well as moving the database, and more. Details about this command can be found here[8].

> **Note**
>
> It is important to remember to consider whether there are third-party backup agents etc. that you will need to install on the new servers. For example, if you are using a third-party backup solution, you might need a new agent and configuration to maintain your database maintenance routines.

One of the key areas that sits outside of the database that is migrated is the backup of database assets. Within Azure VMs there are all of the existing options related to built-in backup and restore capability. However, by moving to Azure VMs it is possible to consider additional capabilities in this space. There are two key options that are available to us:

- Backup to URL

- Azure Backup

Backup to URL is a feature of SQL Server that is part of the built-in backup and restore capabilities. It is relatively seamless to switch from this built-in backup to SMB file sharing to target an Azure Blob storage account. By making use of this we can remove the management of backup locations on Windows file shares or local storage volumes. Additionally, because of the locality of the backup target, the performance considerations around networking that we would have had on-premises are largely mitigated.

Azure Backup for Azure VMs with SQL Server provides a centralized management and reporting pane for your backup infrastructure. This is a native capability within the Azure platform and has a specific SQL Server agent that can be deployed. The data itself is then backed up and stored in Azure Recovery Services vaults.

In my experience, the choice of which to use depends largely on support team models and the larger infrastructure. If the database servers are the only ones being deployed to the cloud at this time, then Backup to URL is my preference as it is still firmly in the realm of the DBA to set up and manage. If there is a larger infrastructure deployment where other IaaS VMs are in play, then I would use Azure Backup as it then provides a single-pane-of-glass view on all backup activity in enterprise deployments on Azure.

Both of these options provide an additional level of protection for SQL Server backups in ransomware scenarios. By having them isolated in another storage platform, it means that in the event that your infrastructure is compromised, you'll be safe in the knowledge that you'll be able to recover.

Migrating databases to the cloud

Now that we understand our workloads, the dependencies, and have all the prerequisite work completed, it is time for us to do the heavy lifting. Now we will look at how we can move our databases to our Azure VMs.

Backup and restore

As of SQL Server 2012 SP1 CU2, it is possible to back up directly from SQL Server to Azure Blob storage. This greatly simplifies the process of migrating databases from SQL Server on-premises systems to those based in the cloud. By using Blob storage, we can remove the need to extend our network to Azure by using secure transfers via the public storage endpoints.

This can be achieved by following the steps shown in *Figure 6.6*:

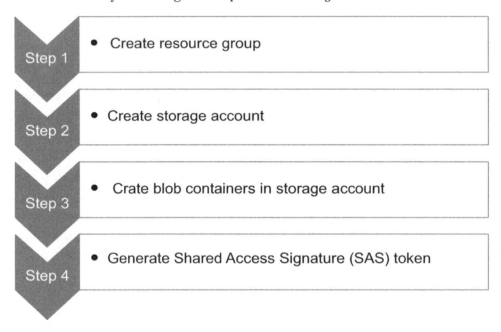

Step 1
- Create resource group

Step 2
- Create storage account

Step 3
- Crate blob containers in storage account

Step 4
- Generate Shared Access Signature (SAS) token

Figure 6.6: Backup and restore procedure in PowerShell

Once we have the Azure storage in place, we can follow the migration process shown in *Figure* 6.7 to move the databases:

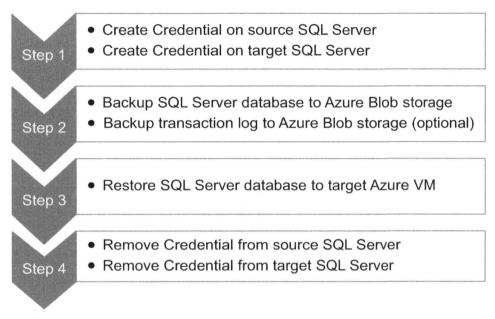

Step 1
- Create Credential on source SQL Server
- Create Credential on target SQL Server

Step 2
- Backup SQL Server database to Azure Blob storage
- Backup transaction log to Azure Blob storage (optional)

Step 3
- Restore SQL Server database to target Azure VM

Step 4
- Remove Credential from source SQL Server
- Remove Credential from target SQL Server

Figure 6.7: Database migration process via backup and restore

> **Note**
>
> It is important to remember when restoring these databases that if the path on the target system is different from the source, you will need to use the MOVE clause in your restore statement and set the file paths that are to be used in the new environment.

Log shipping

Building on the backup and restore methodology, we largely automate the process by leveraging log shipping. It has several additional benefits, which we will cover in this section.

When we want to minimize downtime, we can make use of the log shipping feature in SQL Server to get the source and destination databases aligned. This allows us to minimize the downtime for switching between systems as the final log transfer should be small and quick. All that remains then is to re-point the applications to the new databases.

Log shipping is a tried and tested technique for moving databases between servers and is familiar to many database professionals. This approach also has the benefit of being able to seed multiple availability group replicas ready for migrating to a high-availability configuration.

However, it should be noted that to configure log shipping from on-premises SQL Server systems to Azure VMs, the network will need to be stretched to Azure. This can be done via a service such as ExpressRoute or a site-to-site VPN. By doing this, we are then able to easily make backups locally on-premises before copying them to Azure and restoring them in our VM.

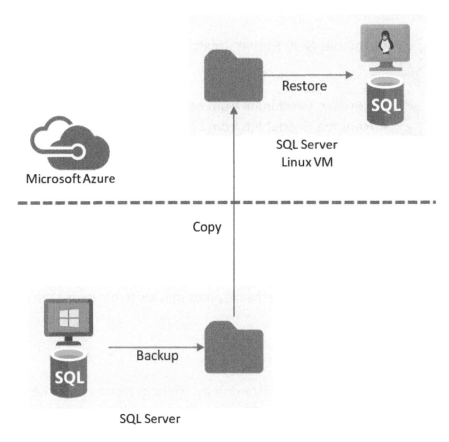

Figure 6.8: Log shipping to Azure topology

As we can see here in Figure 6.8, log shipping is a logical extension and an automated form of backup and restore, which was discussed earlier. It is important to understand how to tune databases for performance by using the built-in backup and restore features of SQL Server, from adjusting the number of backup files, all the way through to the BUFFERCOUNT and MAXTRANSFERSIZE options.

> **Note**
>
> Refer to this documentation for more information on the backup[9] and restore[10] commands.

At this point, we have now understood our workloads, fulfilled prerequisites, and migrated our databases to the cloud. Now we need to think about some of the external factors and considerations around what uses these databases.

Using DMS and DMA to migrate databases

Up until now, we have looked at how to perform the migration of databases manually or with built-in SQL Server capabilities. However, there are other options available in the form of DMA, which we covered earlier, and the Azure DMS.

DMS is a multi-faceted Azure service that can be used to migrate databases from multiple sources to SQL Server in Azure. This service will largely automate and coordinate the migration of database assets from on-premises to the cloud, providing a central migration project dashboard that we can reference. Creating and configuring a migration project is achieved via the Azure portal or PowerShell using the AZ module. This latter ability means we can easily template a migration project, allowing us to perform migrations at a large scale from on-premises to Azure in the event that we have many servers and databases to migrate.

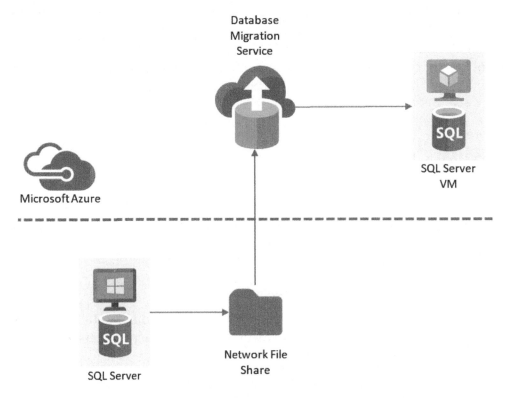

Figure 6.9: DMS migration to Azure VM

When we are moving databases from on-premises to Azure VMs we need to configure some prerequisites within the environment to facilitate this. Notably, these include:

- An on-premises file share that SQL Server can send backup files to. This is an SMB share on port 445.

- An **Active Directory** (**AD**) domain account with permissions to access the share so that DMS can read the data.

DMS will help us automate the migration of databases from one SQL Server system to another by managing the backup and restore process. This is fine when we have one or two SQL servers to move; however, it really comes into its own when managing large-scale migrations effectively. The single pane of glass showing us progress and status is invaluable in coordinating the migration effort.

> **Note**
>
> At the time of writing, SQL Server 2019 in an IaaS VM is not a supported target for DMS. However, this is where DMA can step in and help us achieve our objective.

Previously, we used DMA as an analysis tool to look for blockers in our migration planning. It is also capable of helping us migrate databases from one server to another. Instead of performing an assessment, we create a migration project and specify our source and destination. As before, it is possible to use the command line to automate this process, so again, we can look to start moving larger volumes of databases and the data within them.

Application considerations

While moving the database elements to the cloud, it is important for us to not forget the application layer. For any cloud migration, it is important to consider application details such as its location and its sensitivity to latency. Building hybrid environments where the application remains on-premises while the database is hosted in an Azure VM is entirely plausible. However, if the application cannot handle the increased latency to the database, then you should really consider moving the application into the cloud. By doing so you can eliminate a large portion of the latency.

Beyond that, we need to really understand the way that the application authenticates to the SQL Server database. By lifting and shifting to an Azure VM we can avoid the need to re-engineer an application to support Azure AD for PaaS systems. If we are looking to include SQL Server on Linux as part of our migration, there are some additional steps we need to perform to configure the Linux VM to use AD authentication. This will then allow the users to connect using AD authentication.

> **Note**
>
> Here, you can find the key steps to configure AD authentication[11]. This covers all the steps you need to enable the use of AD user accounts to access database resources.

One big advantage of migrating workloads from traditional on-premises systems to Azure VMs is the ability to scale up and down as needed. By tracking the seasonality in the workloads and usage of our applications, we can plan resource utilization. For example, if we have a financial services system, we can scale the system up for key events like the end of the tax year and later scale it down as required. Likewise, for retail platforms, it is possible to scale up for Black Friday, Christmas, etc. and then scale down during quieter periods.

This increased agility and adaptability provided by cloud platforms ensures that customer needs are met.

Another key element that commonly gets overlooked at the application level is updating the connection drivers used by the application to connect to SQL Server. Over the last few years, Microsoft has resurrected the OLE DB driver, expanded the ODBC driver, and deprecated the SQL Server native client[12]. To support the latest features of high availability and security, we should look to update the connection drivers on the application servers. This also has a benefit in many situations for performance and the stability of the workload placed on the new version of SQL Server.

Congratulations—we have now migrated our application workloads to Azure by lifting and shifting them to IaaS VMs. Hopefully, you can now see that this is not the monster that many make it out to be. Yes, there are pitfalls, but with a clear method and process in place these can be avoided, and a successful migration can be achieved. The only outstanding element that we have not discussed is our reporting capabilities options. In the next section, we will look at the evolution of reporting services and how Power BI fits into the grand scheme of things.

Reporting in the cloud—Power BI

Historically, reporting in SQL Server on-premises has been handled by SQL Server Reporting Services. Over time it has had only minor increments in capability and enhancements. More recently it has been broken out of the main product as a separate download. This has been in tandem with the introduction and rapid expansion in the capability of Power BI Report Server, which is the next generation of on-premises reporting capability.

When making the decision about whether to install **Reporting Services (SSRS)** or **Power BI Report Server (PBIRS)**, or to leverage the Power BI cloud service, it is important to understand the pros and cons of each option. While the Power BI service provides out-of-the-box high availability, huge scalability, and easy access for developers and information consumers, it does come with the downside of ceding control of feature and product update schedules to Microsoft. Contrast this with the VM IaaS deployment, where we have a lot more control over the configuration of the operating system and components that are installed. But this comes with the overhead of us needing to ensure that we get the IaaS design right for the deployment of services in availability zones or groups. We are also responsible for configuring the high availability and disaster recovery pieces of the infrastructure to achieve the uptime SLAs for IaaS VMs.

It is important to remember that the decision here should be primarily driven by which one best meets the requirements of the service we need to deliver. It is very common for technical teams to fall back on what they know and pick a technology out of comfort. Once the requirements are established and evaluated, the key driver becomes supportability. Putting a system in place where we have minimized our support overhead will put the long-term stability of the solution in a much better place.

PBIRS has several of the capabilities of the cloud-borne Power BI service but at the same time lacks the richness of its bigger sibling. One of the key drawbacks is that it is a software service that needs to be installed and managed within environments. On the other hand, the Power BI cloud services act as a SaaS/PaaS model. This means that we do not have to worry about updates and upgrades throughout the product lifecycle.

Figure 6.10 illustrates a basic PBIRS deployment scenario. We have the report server on-premises and it can connect to local and cloud-based relational engines.

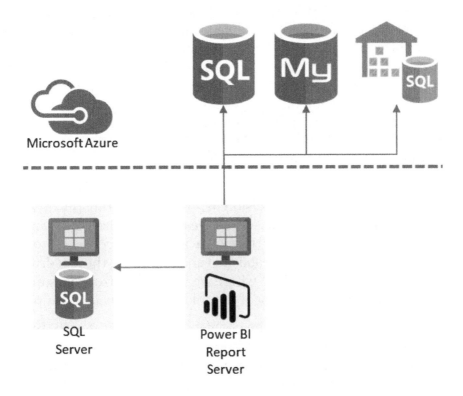

Figure 6.10: Power BI Report Server on-premises

When we contrast this with the cloud deployment model, we can clearly observe the increase in scalability and capability. This is illustrated in *Figure* 6.10. Here we can leverage the capabilities of Power BI to connect to a multitude of services beyond relational engines, including web services and APIs for many common cloud applications. Here's a complete <u>list</u>[13] of data sources that Power BI supports for datasets. We can also deploy Power BI gateways close to our data sources if we are operating them on-premises for hybrid scenarios or in multi-cloud environments.

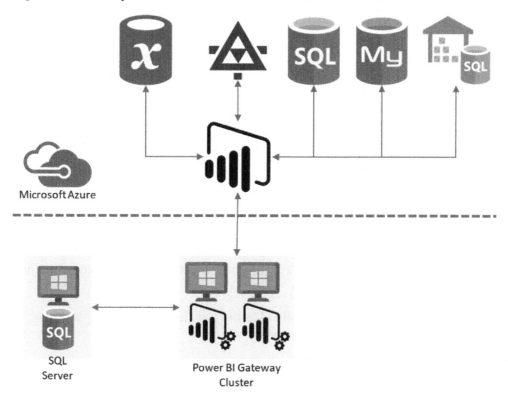

Figure 6.11: Power BI Cloud service deployment

When we make the jump from on-premises or IaaS hosted PBIRS to the PaaS Power BI service, the whole way we think about and manage the platform will change—moving from hierarchical folder structures with prescribed permission sets to workspaces. These workspaces can be managed or personal, with users able to collaborate and share content easily. As such, when moving to the PaaS-based solution, putting the effort in to get the governance in place early on is very important.

After we have made the decision to move to the Power BI service, we need to think about what types of reports our users consume. Do we simply lift the paginated reporting from PBIRS and place it in the service in workspaces that mimic our folder structures? Or, do we look to realize the capabilities available and leverage more dynamic and diverse dashboarding and reporting capabilities? The former is more akin to the lift and shift approach from earlier when we looked at the database elements. If we want to take the latter option, we can lift and shift the native reports (`.pbix` file) deployed on-premises to the cloud service. But to realize the true capabilities and value for our users, we will need to look at re-working some of the reports and dashboards. However, this can be done incrementally so, while not ideal, there are no blockers preventing us from rapidly using the Power BI service.

As with the migration of SQL Server databases, which we covered earlier, there are other supporting artifacts that we need to migrate too. These include, but are not limited to, email subscriptions, custom visualizations, security configurations, and bookmarks. Depending on the complexity of your deployments, the volume and scale of these elements will vary.

Other key elements that we need to consider during the deployment and utilization of the Power BI service are:

- Security and integration with Azure AD.
- Pro versus premium tiers.
- Monitoring and managing the Power BI service.
- Licensing and user management.

By combining all these elements and migrating our on-premises PBIRS to cloud services, we can build and manage a comprehensive analytics platform based in the cloud.

There are several Microsoft whitepapers[14] covering planning, deployment, and the management of Power BI. These will help you plan and execute a successful deployment and migration to Power BI from on-premises systems. This content is invaluable in creating strong governance and an operating model to ensure data security and availability for our users.

ETL in the cloud

While our reporting platforms can connect to most data sources and allow us to perform operational reporting on OLTP platforms, when it comes to strategic reporting it is very common for us to need to report from a centralized repository of data that brings many sources into one place. On-premises the tool of choice is **SQL Server Integration Services (SSIS)**. In the cloud we have a couple of other options available to us.

The shortest route to getting our ETL working in Azure is to deploy SSIS to an Azure VM and then deploy the SSIS project to this server. This mirrors our on-premises configurations and requires a very limited re-work of SSIS projects.

However, by moving our workloads to Azure, we also get the option of **Azure Data Factory (ADF)** to run our ETL workloads. There are two options here: one being to rewrite our SSIS ETL processes as ADF pipelines and the other being to host our SSIS packages in Azure and execute them through the Azure SSIS integration runtime[15].

This latter option allows us to build a hybrid ETL solution where we can leverage our existing on-premises SSIS projects with very little change, as well as allowing the use of ADF for new development and controlled migration to the PaaS ADF service.

We have the ability to take a complete on-premises data platform ecosystem and run it in Azure IaaS VMs. Alternatively, with the Power BI service and ADF, we have the ability to migrate our data stores and at the same time leverage the PaaS capabilities of Azure to streamline our operations.

Summary

As we worked through this chapter, we had a look at how we can accelerate our adoption of cloud data platforms with SQL Server and Power BI. Microsoft recommends that the fastest way to realize the value of SQL Server in Azure is to lift and shift. This has many benefits from a familiarity perspective, meaning the removal of many barriers.

We looked at how to prepare for and then execute a migration to SQL Server on Azure VMs, highlighting several great free tools that are available to ensure a successful migration. Here's a step-by-step guide on how to migrate from on-premises SQL Server to SQL Server on Azure VMs[16].

But this is only the beginning. Once we have got our databases into Azure, we can then look to take the next steps and move from an IaaS solution to PaaS. Many of the tools and techniques we have covered here still apply. There are more considerations on the application front, especially around authentication. But making the step to move to Azure SQL Managed Instance or Azure SQL Database is much smaller once we are there.

In the next chapter, we will be looking at how to use SQL Server in Azure VMs for building hybrid data platforms, looking at some of the key considerations of data replication, availability, and disaster recovery scenarios.

Chapter links

1. https://bit.ly/36uDXcF
2. https://bit.ly/2ZzTqXu
3. https://bit.ly/2AdSLjF
4. https://bit.ly/2X1ZJkT
5. https://bit.ly/2TBpsyE
6. https://bit.ly/2zlUtzJ
7. https://bit.ly/3gpoLSP
8. https://bit.ly/2TG9Zgz
9. https://bit.ly/2X3uMg0
10. https://bit.ly/2ZDzh2S
11. https://bit.ly/3eirDPl
12. https://bit.ly/2AWXSoN
13. https://bit.ly/2B2sGVn
14. https://bit.ly/3d7OcWQ
15. https://bit.ly/3bXZtYh
16. https://bit.ly/2LXJM93

Hybrid scenarios (Microsoft SQL IaaS)

By Randolph West

This final chapter builds on what you've learned in the previous six chapters and discusses the various ways in which you can develop a hybrid environment, leveraging Azure services to complement your on-premises SQL Server environment.

We will explore several Azure licensing and technical offerings, including Azure Hybrid Benefit and Backup to URL. We will discuss the basic principles of disaster recovery, and then provide use cases for Azure VMs running SQL Server on Windows and Linux, known as **infrastructure as a service (IaaS)**. We'll also cover the ways you can keep a workload in sync between your on-premises and Azure environments, and how these relate back to scalability, migration, and disaster recovery scenarios. We'll finish with a summary of the chapter.

What is Azure Hybrid Benefit?

The majority of organizations that use SQL Server do so in an on-premises environment with SQL Server Standard or Enterprise editions running on physical or virtual machines in a datacenter.

With the advent of Azure cloud services, you might be tempted to explore this new world but feel that your investment in on-premises licensing is keeping you from doing so. With Azure Hybrid Benefit, you can make use of Azure services at a reduced cost, provided you have Software Assurance or an equivalent subscription license.

> **Note**
>
> Software Assurance is not a license. Instead, you can think of it as an additional benefit that either comes with the license or something that you purchase as an add-on as part of your volume licensing deal. You can read the Software Assurance Frequently Asked Questions at this Microsoft documentation[1].

This unique offering gives you the opportunity to build scalable, highly available, and disaster-resilient solutions for your organization, without needing to invest in ongoing hardware maintenance, saving you money and resources in the process. Azure Hybrid Benefit even gives you a free asynchronous disaster recovery replica in Azure if you use availability groups. We discuss this scenario later in the chapter, in the *Use cases for SQL Server on Azure VMs* section.

There are three key scenarios that a hybrid SQL Server infrastructure provides:

- A fully redundant disaster recovery environment in the cloud

- Secure offsite backups to Azure Storage using Backup to URL

- Read scalability across regions with readable secondaries and replication

> **Note**
>
> You can read more about Azure Hybrid Benefit in the Frequently Asked Questions available at this Microsoft documentation[2].

Now that we've covered Azure Hybrid Benefit, let's dig into the basics of disaster recovery.

What is disaster recovery?

When disaster strikes, your job as a SQL Server database administrator is to ensure business continuity by recovering the data estate to a previously known good state, in as short a time as possible. Disaster recovery is your organization's insurance policy and relies on support from the organization as well.

> **Note**
>
> A disaster is any event that causes an unplanned interruption in business continuity through unrecoverable failure.

At all times, remember that high availability is not disaster recovery. You're considering what happens when high availability could also fail, even if it makes use of the same underlying technology.

A good disaster recovery plan starts with a healthy database. You can make use of the maintenance features inside SQL Server, including maintenance plans with SQL Agent inside SQL Server Management Studio and Azure Data Studio, PowerShell cmdlets, from the command line using SQLCMD, and those provided by third parties, to keep your databases in good health.

Take native backup—always. If you need to perform point-in-time recovery for any reason, then you need to know about the recovery models SQL Server offers (full, bulk-logged, and simple). This, in turn, will help you learn the difference between full, differential, and transaction log backups. Practice backing up and restoring databases, including system databases. Learn when you would need to do a copy-only backup, and how they affect differential backups.

You can back up your data estate using Azure Storage as a target, which we discuss in the *Backing up databases to a URL* section later in this chapter.

Once you've backed up your database, the only way to prove that your backups can be recovered to a previously known good state is to continuously test that backup. The best way to test a SQL Server backup is by restoring it and running a data consistency check with the **DBCC CHECKDB Transact-SQL (T-SQL)** command against it. No other method can provide the same peace of mind. If your organization uses file-system-level backups, make sure you also have SQL Server native backups close by.

Finally, and especially for the purposes of this chapter, you need to make sure that your verified backups are copied securely offsite (remember that most databases contain sensitive data) so that if something happens to your on-premises environment, you can get those backups and restore them as fast as possible. Encrypt your backups, even if you don't encrypt your database, and you should encrypt your database.

> **Note**
>
> Please refer to Microsoft Docs[3] for information about disaster recovery with Azure SQL Database.

Disaster recovery is defined by your organization in a **service-level agreement** or **SLA**. This document may be part of a larger business continuity plan, and it should comprise at least two main points:

- **Recovery point objective (RPO)**
- **Recovery time objective (RTO)**

We'll talk about each of these in the following sections and give you a refresher on **Accelerated database recovery (ADR)**.

Recovery point objective

When disaster strikes, your goal is to ensure **minimal data loss**, which is usually measured in seconds or minutes. The RPO defines how much data your organization is prepared to lose should something go wrong. For example, your office building suffers an electrical fire and the server room burns down. If your SLA states that up to 15 minutes of data loss is acceptable to the business, and you have a plan in place to keep your disaster recovery site in sync within a 5- to 10-minute window, then you are probably going to be within the requirements of the SLA.

Even with a hybrid setup, you need to make sure you are taking regular database backups and the backup frequency should be at least one half of the data loss window to allow time for backups to find their way to a secure offsite location. For example, say your RPO is 15 minutes. If your transaction log backups are set to a 15-minute interval and a disaster occurs after a backup has taken place but before those backups were copied securely offsite, your data loss will be greater than 15 minutes, taking you outside the requirements of the SLA.

Recovery time objective

After a disaster has taken down your data estate, the RTO dictates how quickly you need to bring up the environment again, which is usually measured in minutes or hours. It doesn't matter how you achieve this, and generally, budget and resource constraints will define the most cost-effective manner to do so. A major consideration is how you prioritize data recovery. You may have a lot of archive data that is infrequently queried, which does not need to be brought online immediately. Your SLA must define terms for this process as well.

Again, even in a hybrid environment, you must ensure that your database backups can be restored as quickly as possible. It is feasible that a failover could itself fail, so think of your backups as a last resort that must always work. Use a combination of full, differential, and transaction log backups to reduce this recovery window.

> **Note**
>
> You can think of transaction log backups as incremental backups, whereas differential backups can be used as a shortcut to avoid having to restore each transaction log in sequence. Both differential and log backups can be combined with the last full backup to bring your database online.

The DR plan should also make allowances for network-related failover, including DNS propagation. We recommend that you set your **time-to-live** (**TTL**) values to between 5 and 15 minutes, ensuring a faster failover if an IP address must be updated.

Accelerated database recovery

ADR is a new feature introduced in SQL Server 2019 that works at the database level. It is not enabled by default because it changes how SQL Server uses the data file. However, its benefits are most apparent during crash recovery, which happens whenever a database is brought online, restored from backup, or when SQL Server starts up. Instead of using the transaction log file or TempDB, ADR makes use of the main data file for keeping track of transaction state and can dramatically reduce rollback times for long-running transactions.

> **Note**
>
> ADR is also available in Azure SQL Database.

ADR increases data file usage, but the trade-off is improved recovery times, which benefits disaster recovery.

Now let's address some licensing benefits around disaster recovery before getting into the next section.

How does licensing influence disaster recovery?

Under the terms of Azure Hybrid Benefit, you are allowed two failover instances for disaster recovery for each primary workload, provided that one of these is located on an Azure VM. For the second server, the only condition is that it must be dedicated for your use. It can be hosted on-premises or on an Azure VM.

The benefit is calculated based on the number of cores licensed for your primary workload and depends on whether your primary workload is located on-premises or on an Azure VM. In both cases, you get one free passive core for high availability / disaster recovery and one free passive core for disaster recovery (asynchronous commit only). If your primary workload is on-premises, you also get one free passive core for DR on SQL Server on an Azure VM (asynchronous commit only).

Note that these benefits apply exclusively to the SQL Server core license. You are responsible for Azure-related costs including Azure Storage, compute, and networking. If you have questions, you can always speak to your Microsoft licensing specialist to determine what rights you have as per your enterprise agreement.

> **Note**
>
> Read more about these benefits on the SQL Server blog[4]. Additional information on SQL service licensing and disaster recovery benefits can be found at this Microsoft blog[5].

In the next section, we'll look at an easy way to get database backups copied securely off-site.

Backing up databases to a URL

Starting with SQL Server 2012 with Service Pack 1 CU 2, you can back up your SQL Server database to an Azure Storage account, also known as **Backup to URL**. Since SQL Server 2016, you can even back up your database to a URL simultaneously with a regular on-premises backup. This allows you to have both a local and secure offsite backup of your databases from one backup command.

> **Note**
>
> Backup to URL can be executed from the **SQL Server Management Studio (SSMS)** backup wizard, T-SQL, SQL **Server Management Objects (SMO)**, and PowerShell cmdlets, including third-party PowerShell modules. For more about this feature, visit Microsoft Docs[6].

Backup to URL gives you the peace of mind that when disaster strikes, you will be able to access your backups almost immediately. With a solid disaster recovery plan, you can have a standby Azure VM restoring those backups on a regular schedule (see the *As a backup-restore target* section later in this chapter). You can even configure your Azure Storage account to have a geo-redundant replica in a secondary region, which provides even greater redundancy.

How to back up to a URL

To back up to a URL, you first need an Azure Storage account and container. The container must be set to **private** (the default) to ensure authorized access only. Then, you will create a **Shared Access Signature (SAS)** in the Azure portal to allow access to that container. Inside SQL Server, you will create a credential that uses that SAS when performing the actual backup and restore tasks.

Once you have a Storage account, container, and credential, you will then choose a blob type. With the SAS credential, you will choose the **block blob** type. If your backup exceeds 200 GB in size, you can use backup compression and striping to back up larger databases.

To get the best performance out of Backup to URL, Microsoft recommends using the following arguments:

- `BLOCKSIZE = 65536` (65,536)

- `MAXTRANSFERSIZE = 4194304` (4,194,304)

Now let's dive into the main section of this chapter, namely SQL Server on Azure VMs.

Use cases for SQL Server on Azure VMs

This section will demonstrate ways to leverage Azure VMs running SQL Server in a hybrid scenario, for scalability, migration, and disaster recovery.

As we mentioned previously in this book, SQL Server 2019 runs on both Windows and Linux, with almost complete feature parity between the two operating systems. You will access SQL Server on Linux using the same set of tools, and many of the common tools for scalability, migration, and disaster recovery, including backup and restore, log shipping, consistency checks, transactional replication, and so on. Even availability groups can be leveraged in a hybrid environment using an asynchronous commit.

The following sections present three sample use cases for Azure VMs in a hybrid environment:

- As a backup-restore target

- As an availability group replica

- As a transactional replication subscriber

The important thing to remember here is that a technology or feature doesn't have a single use. Each sample can be used for read scalability, database migration, and disaster recovery. It is worth mentioning that these methods should be considered read-only where your on-premises instance is the single version of the truth. In other words, the data flows in one direction.

As a backup-restore target

You can keep a standby server in sync by restoring transaction log backups for one or more databases on a regular schedule. There is already a feature built into SQL Server called **Log Shipping**, which uses a shared network location to transfer regular backups between servers on the same network and restores them on a schedule. Using those same principles, you can achieve the same outcomes with an Azure VM as your target. You just need to ensure that the transaction log backups are securely copied to a shared location, where the standby server is waiting to pick them up and restore them in the correct order.

When leveraging an Azure VM as a standby server, you can back up your transaction logs to a URL, which stores those files in your Azure Storage account. After a successful backup, you can generate a T-SQL restore script from the backup history in your on-premises **msdb** database, and transfer that script to the same storage account (using a synchronization tool like AzCopy inside a PowerShell script, for example). On the Azure VM side, you will retrieve the T-SQL script on a schedule, also using AzCopy, and then run the script that restores the databases directly from Azure Storage.

Backup-restore is a cost-effective way to get your on-premises database to synchronize with a standby server, and it works on any edition of SQL Server. Keep in mind that all connections to the restoring database will be dropped when the next transaction log is restored.

As an availability group replica

While Always On availability groups provide a high availability solution to ensure that your environment has minimal downtime, they are not best suited to sharing a workload between your on-premises Windows instance and an offsite replica (running on Windows or Linux). For this, you would use either one or more clusterless replicas, or an entire distributed availability group, in combination with ADR.

A clusterless availability group replica is kept mostly in sync using the asynchronous commit mode. Depending on network latency and throughput, you may find that your VM is able to keep up with your workload. If it falls behind, you can monitor this according to the requirements of your SLA.

If you want more redundancy, with the understanding that this is still not a high availability scenario, you can build a distributed availability group with multiple nodes in its own availability group. Your on-premises primary replica will send data to the distributed availability group via the forwarder, which is the primary replica in the distributed availability group.

> **Note**
>
> You can read more about distributed availability groups from Microsoft Docs[7].

As a transactional replication subscriber

A third option is transactional replication, which shares your workload between your on-premises environment and SQL Server on an Azure VM at the individual table level. You set up a publication and distributor on-premises and have the Azure VM as the subscriber. You define which tables will be synchronized, and let replication keep your databases in sync. This flexibility allows you to point end-users to the new Azure VM for read-only purposes, thereby scaling out your environment without the expense of scaling up your on-premises infrastructure.

The two types of SQL Server replication supported by both Windows and Linux are:

- **Transactional replication**: Best for servers that are in constant communication and need to deliver data downstream all the time. Data flows from the primary database (distributor) to one or more secondary databases (subscribers) elsewhere. Remember to ensure that your replication environment is appropriately licensed.

- **Snapshot replication**: Used for creating the original snapshot before transactional replication takes over and is useful when you need to perform a refresh of the entire data set.

> **Note**
>
> **Peer-to-peer transactional replication** and **merge replication** are not supported on Linux. If you require bi-directional sync, your publisher and subscriber should both use Windows Server. For more information about SQL Server replication, refer to this Microsoft documentation[8].

SQL Server on Linux supports **Active Directory** (**AD**) authentication, which means that AD is supported with replication. Provided that the appropriate network ports are open on the Azure VM firewall (and associated Azure Network Security Group), SQL Server does not care whether the underlying operating system is Windows or Linux.

> **Note**
>
> You can follow a detailed walkthrough for setting up replication on Linux at Microsoft Docs[9].

Hybrid scenarios

Taking what you've learned from this book up to this point, including the earlier parts of this chapter, you can start to build a picture in your mind of how SQL Server on an Azure VM can help you with creating a hybrid solution at any scale.

Whether for read-scalability, migration, or disaster recovery, you can run your on-premises SQL Server instance on Windows Server and have the same workload in SQL Server on an Azure VM at the same time.

Keep in mind that with Software Assurance, the free passive SQL Server replica can be used to synchronize with your primary replica (using asynchronous commit and manual failover) and run these maintenance operations:

- Database consistency checks
- Full and transaction log backups
- Monitoring resource usage data

Additionally, you can run disaster recovery testing every 90 days with primary and disaster recovery replicas running simultaneously for brief periods.

Should you desire additional operations for your replica, including synchronous commit and automatic failover, your secondary disaster recovery replica must be appropriately licensed. With Azure Hybrid Benefit, for example, you can have a failover secondary in synchronous commit mode with automatic failover.

Scenario 1: Read scale workloads

You have an on-premises SQL Server database located on the West Coast of the United States and customers all over the world: Johannesburg, Toronto, Seoul, London, and Sydney. Your customers want to query the database for reporting and analysis, but they are complaining about performance introduced by network latency.

As the DBA, you notice performance issues on your production server due to locking and blocking. If you could somehow offload that workload into a read-only copy of the data that is closer to them, you could free up resources in your production environment.

This is a typical read-scale scenario. Using the customer distribution above, you can place Azure VMs running SQL Server on Windows or Linux in the regions closest to your customers, for example, West US (for customers closest to you), South Africa North, Canada Central, Korea South, UK South, and Australia East.

> **Note**
>
> You can see all available Azure regions at this Microsoft documentation[10].

You can use all three use cases described in the previous section, with the following provisions:

- **Backup-restore**: Customers will be unable to connect during a restore process. To ensure a better user experience, use ADR to speed up the rollback portion of the restore process, and add retry logic to your application code to work around disconnections.

- **Readable secondary replica**: Queries on the readable secondary might impact performance on the primary replica and vice versa.

- **Transactional replication**: Each table that will be replicated must be configured individually, but the benefit is that you only need to replicate the tables that are needed for read scalability. You are trading more configuration up front at the publisher and subscriber level, against more granular access to the data for your customers.

In fact, you could use a combination of all three, because customers in each location may have different requirements. You have a lot of flexibility with read-scale workloads, and a hybrid Azure infrastructure is well-suited to this scenario.

Scenario 2: Migrating a workload

Assume you are preparing to lift and shift an on-premises database or instance to SQL Server on an Azure VM.

You may think that the easiest way to move a database from one server to another is by detaching the database, moving or copying the files to the new server, and reattaching them; however, if you move the files, there is no way to roll back if the reattach fails. So always (always!) ensure that you have a valid, tested, full SQL Server backup of your database.

> **Note**
>
> Read more about detaching and attaching databases at Microsoft Docs[11].

Irrespective of whether you restore or reattach the database, and depending on the version of SQL Server at the other end, the internal system tables will be automatically upgraded to that build. Once the database is brought online again, you can continue using it.

Both backup-restore and detach-attach require downtime. With techniques like log shipping, clusterless availability groups, or even replication, you can reduce the downtime dramatically.

We'll cover ways to keep your primary database up for as long as possible before cutting over to the new server, which also makes it easier to roll back if something goes wrong.

In this scenario, you can use two of the use cases described in the previous section, with the following provisions:

- **Backup-restore**: This use case requires some downtime while failing over to the new environment. You should plan a maintenance window to cut over to the new VM and allow time for data consistency checks, DNS migrations (see the note on TTL earlier in this chapter), and updating application connection strings.

- **Distributed availability group**: Instead of having a read-only replica, you will create a new distributed availability group. During a planned maintenance window, you will wait for any outstanding transactions to synchronize to the new availability group, then update application connection strings to point to the new availability group listener. The advantage is that the new availability group is already highly available.

SQL Server replication is not recommended for a migration scenario. While peer-to-peer transactional replication and merge replication both offer bi-directional synchronization, those benefits are outweighed by the complexity and lack of flexibility at a database level.

> **Note**
>
> You can use Azure Site Recovery for migrating a workload to Azure, but you must plan your failover during a maintenance window, and your target operating system must match your on-premises environment. You cannot use Azure Site Recovery to fail over from Windows to Linux. Visit Microsoft Docs[12] for more information.

Scenario 3: Disaster recovery

Your disaster recovery plan is defined by the organization's SLA, and that SLA is guided by how much money the organization is prepared to spend, along with the resources it has available to perform the recovery. Your solution might be as simple as a PowerShell script that creates an Azure VM on the fly and restores the databases from a shared location like Azure Storage.

Or, as we've demonstrated in this chapter, you can dramatically reduce the RTO by having a standby server ready to go, which is kept as up to date as possible in an automated way, using backup-restore, clusterless replicas, distributed availability groups, Azure Site Recovery, or even replication. This way, when disaster strikes, you have much less to worry about. Even so, it is always worthwhile having a Plan B (where B stands for "backups"), where you can restore your databases no matter what.

> **Note**
>
> As mentioned throughout this chapter, using the same features to achieve different goals is common in SQL Server.

Should an unplanned failure occur and the potential data loss due to latency is within your RPO, you can be up and running on your new server in seconds or minutes, especially with ADR enabled.

In this final scenario, you can use two of the use cases described in the previous section, plus a bonus option:

- **Backup-restore**: This can be used to keep your database in sync in the event of a disaster. Whether you use the built-in log shipping configuration or roll your own solution, SQL Server can keep restoring your log backups on a regular schedule, such that the amount of data you can afford to lose (defined by the RPO) is protected by the frequency of log file restores on your Azure Linux VM. With cost-effective monitoring in place, you can keep track of how far behind the target database is. With ADR, recovery time is dramatically reduced, especially for long-running transactions.

- **Clusterless replica or distributed availability group**: You can reduce your RTO by having either a distributed availability group or a clusterless availability group replica on standby, keeping in mind that RPO. You can also keep track of the synchronization state using built-in monitoring features of SQL Server. You can even use the same underlying SQL Server instance that you configured for read scalability. The server can be readily failed over to a primary availability group in the event of a disaster. In the case of a distributed availability group, you are failing over to a highly available solution.

- **Azure Site Recovery**: This is primarily a disaster recovery solution, where physical servers and VMs running on VMware or Hyper-V can be replicated to a secondary site. Applications and workloads supported include SQL Server. This is the recommended disaster recovery solution for a hybrid infrastructure.

> **Note**
>
> You can read more about SQL Server support in Azure Site Recovery on Microsoft Docs[13].

Summary

Building a hybrid infrastructure leveraging Azure services that support your on-premises environment is straightforward. Whether for scalability, migration, or disaster recovery, an Azure VM running SQL Server works the same way on Windows or Linux, with the benefit of reduced OS licensing costs when considering Linux. With software assurance and Azure Hybrid Benefit, you can reduce your costs further by leveraging this Azure VM as a standby server using some of the techniques discussed in this chapter.

Chapter links

1. https://bit.ly/36vt762
2. https://bit.ly/3d3W9wg
3. https://bit.ly/2XyofcB
4. https://bit.ly/2yy72HH
5. https://bit.ly/2ZyWvqN
6. https://bit.ly/2LUVUaW
7. https://bit.ly/2Ac3sna
8. https://bit.ly/3bWtqbm
9. https://bit.ly/3c1mSrV
10. https://bit.ly/2WZvf2S
11. https://bit.ly/2LXCWQV
12. https://bit.ly/3gkxkOx
13. https://bit.ly/2XvLl3u

Appendix A

SQL Server Configuration with OLTP

In **online transaction processing (OLTP)** workloads, much of the workload is made up of thousands of low-cost queries. When a query comes into SQL Server, the query optimizer creates an execution plan for that query. The execution plan determines how SQL Server will gather the data, such as which tables and indexes to use. It also assigns an overall cost to each query regarding how much effort it will take to complete that query. That query cost is known as the estimated subtree cost. SQL Server has a method by which it determines which queries it will process in parallel and those that it won't. It determines this by the **cost threshold for parallelism (CTP)** value.

By default, the initial value set for CTP is 5. It is widely accepted that this value is too low for most environments and should be increased to a higher value. The appropriate value will ultimately depend on the workload running on the server. This can be determined by looking at the cached plans of your parallel queries. You'll be able to review which queries are being executed in a parallel manner and their execution counts. By reviewing these, you'll be able to determine what the value should be to force the more trivial plans to not be parallel and to allow your more expensive queries to take advantage of parallelism. Usually, a starting value of between 25 and 50 is an acceptable range. However, there are organizations that need a much higher value.

OLTP workloads can also generate many single-use ad hoc queries. Single-use ad hoc queries can bloat the plan cache, wasting valuable memory. Enabling the optimize for ad hoc workloads option on the server will store a small compiled plan stub rather than the entire execution plan. If the ad hoc query is executed again, the full compiled plan will be stored in the plan cache. In heavy OLTP workload environments where the optimize for ad hoc workloads option is not enabled, you could see several GB of memory wasted storing single-use plans.

Auto growth sizes for user databases should be configured for both data and log files to a fixed size rather than percentage. If a database is growing rapidly, you don't want it to have to grow by many small increments. Setting an appropriate initial size for the data and log file with a fixed size auto growth value helps prevent SQL Server from having to pause, grow, and resume each time the data and log files need to grow. Although the process is very quick, an excessive number of growths per day can impact the server.

Backup compression can also cut down on the time it takes to back up and restore a database. Backup compression does consume additional CPU, but most large backups (full and differential) are scheduled to run during non-peak times. The benefit of faster backups and restores usually negates any additional CPU load. Compressed backups also take up less space, which can save on storage costs. Backup compression can be enabled for the entire instance or specified at the time the backup is made using the **WITH COMPRESSION** command.

SQL Server databases encounter index fragmentation due to updates, inserts, and deletes. Index fragmentation is the result of data being moved around, leaving behind free space in the data pages. When SQL Server reads those pages, the empty space is also retrieved, which wastes space in the buffer pool. SQL Server must also scan and read more pages than it should if the data wasn't as fragmented.

For example, since SQL Server stores data on 8k pages, if those pages were 100% full and a table had 100 pages, after a number of updates, inserts, and deletes, those tables would become fragmented. That same data may now be spread across 125 pages that are only 75% fragmented. SQL Server would now have to scan 125 pages of data rather than the original 100. To control index fragmentation, regular index maintenance needs to be performed.

Index fragmentation can be remediated by performing index rebuilds or reorganizations. Index rebuild drops and recreates the index. For indexes with high levels of fragmentation, this could be a less expensive operation than reorganizing. Index rebuilds can be an online operation for Enterprise Edition, but it is an offline operation for Standard Edition, meaning that on Standard Edition, when the index is rebuilt, it is not available.

An index reorganization uses fewer system resources and is an online operation, making it the preferred method for Standard Edition instances for 24/7 workloads. You can automate index fragmentation by using SQL Server database maintenance plans, which can be configured to either rebuild or reorganize indexes based on fragmentation levels. There are also third-party tools and scripts available to address index fragmentation.

An index rebuild will update statistics for the table or indexed view. Statistics are used by the query optimizer to generate execution plans. Having up-to-date statistics can greatly improve query performance. By default, auto-update statistics are enabled on the SQL Server instance. Depending on the workload and size of the environment, statistics may not be updated frequently enough to ensure that queries are being compiled with current statistics. Creating an automated process to manually update statistics on a regular basis can ensure that the query optimizer has more up-to-date statistics to create more efficient plans. This is a problem that many organizations don't know they have, and the fix is simple.

Organizations relying on index fragmentation processes that rebuild and reorganize indexes based on fragmentation levels could be an issue. This is because index rebuild statement updates statistics, but an index reorganization does not. If an index in that environment never reaches the threshold to be rebuilt, the statistics would not be updated until they hit the auto update statistics threshold, which may not be soon enough. Organizations that are using a process to reorganize or rebuild indexes based on fragmentation levels should also have a process in place to update statistics.

Use the latest compatibility level for your databases when possible. This will ensure that you are utilizing the most up-to-date cardinality estimator for your version of SQL Server, as well as any features or functionality that are tied to that compatibility level. Compatibility levels are not automatically adjusted when you upgrade from one version to another for user databases. You must make this change manually.

Database corruption can happen. SQL Server includes a set of database console commands for administration tasks. **DBCC CHECKDB** is one of those commands. **DBCC CHECKDB** should be run periodically to check for any corruption that may have occurred. If a non-clustered index gets corrupted, you can simply drop and recreate the index. However, if the corruption is on a heap, clustered index, or system table, a database restore is most likely the only fix. You want to ensure that your backup retention and the interval for running **DBCC CHECKDB** provides adequate backups for you to be able to respond and restore data with any loss.

For example, imagine a scenario in which you only run **DBCC CHECKDB** once a week. You've also opted for just 7 days of backup retention. If the database were to be corrupted just after a **DBCC CHECKDB** was scheduled, then there would be almost 7 days to wait before the check would run again. This means that due to the fact you only have 7 days of backup retention, almost the entirety of your backup would be corrupted before another scan would detect the corruption and allow you to respond. A good policy would be weekly checks and at least 30 days of backup retention. Unfortunately, too many organizations find out too late that they have a problem and encounter data loss.

Index

About

All major keywords used in this book are captured alphabetically in this section. Each one is accompanied by the page number of where they appear.

Made in the USA
Middletown, DE
30 June 2020